A DICTIONARY OF
MILITARY
Quotations

A DICTIONARY OF
MILITARY
Quotations

Compiled by
TREVOR ROYLE

SIMON & SCHUSTER

New York London Toronto Sydney Tokyo Singapore

First published 1989
by Routledge
11 New Fetter Lane, London EC4P 4EE

and in the USA by
Academic Reference Division
Simon & Schuster
15 Columbus Circle
New York, NY 10023

Printed in Great Britain

Library of Congress Cataloging-in-Publication Data

Royle, Trevor.
A dictionary of military quotations / compiled by Trevor Royle. —
1st American ed.
p.   cm.
ISBN 0-13-210113-0
1. Military art and science—Quotations, maxims, etc.   I. Title.
U102.R784   1990
355′.003—dc20   90-34034
                                                              CIP

CONTENTS

PREFACE

The object of this dictionary of quotations is to offer the reader a set of keen insights into the various aspects of military life. Many of these are to be found in the sayings of the great commanders of history who led victorious armies in battle. Naturally enough, they have had incisive things to say not only about the nature of battle but also about their soldiers' reactions to the stress of combat. In choosing their most memorable quotations I have tried to shed light on the condition of the soldier in the heat of battle and in the quiet of the barracks. Concepts as different as morale and comradeship, fear and courage, discipline and cowardice, recur again and again simply because they are such vital ingredients of the soldier's craft. I make no excuse for emphasising such factors and the careful reader will discover that there is often a commonality of thinking on military matters which is shared by the great commanders.

This is not to say that I have ignored the musings of the ordinary soldier. On the contrary the annals of warfare are well stocked with the diaries, letters and reminiscences of simple men who have been transformed by their experience of battle. The Peninsular War offered particularly rich pickings mainly because the volunteers who fought believed that they were engaged in a great cause: the same holds true for the later Crimean War but for the opposite reasons. By the time that the mass armies came into being to fight the great world wars of this century raw anger and outraged pity had entered the vocabularies of the private soldiers. Here I have tried to strike a balance between the literary insights of the soldier–poets and the thoughts of the men who committed their words to private papers. Some quotations are longer than others and may verge on anecdotage: they have been included at length because any foreshortening would have removed their pithiness or originality.

Because the book is intended for an English-speaking readership the selection of quotations is principally British and American with significant forays further afield. The time span begins in Classical times to embrace the histories of Xenophon, Herodotus, Thucydides, Caesar and others and continues until the present day. Some subjects are more comprehensive than others and herein lies a word of caution. The length of a subject's entry does not always signify its historical importance. Certain commanders, for example, may have been more witty than others, or may have written more fully about their lives and careers or been fortunate in their choice of biographer. Others may have had less to say about their own lives or their attitudes to war and peace. Similarly, some campaigns have been more faithfully recorded than others: a whole book of quotations could be given over to the Peninsular War alone.

For ease of reference the dictionary has been divided into five sections. The first, *Captains and Kings*, supplies the sayings of the world's greatest commanders and military theorists. Old chestnuts have been included as well as less familiar quotations which seemed to me to be illustrative of the subject's character or personality. If a particular quotation cannot be found under the main subject entry reference should be made to the indexes. In many cases it was more sensible to include a commander's remark about 'morale', for instance, under that heading instead of under his own name.

The second section deals with some of the great battles and wars and here again the emphasis is on British and American conflicts. To have included quotations from all the wars of antiquity or the warfare of the Middle Ages and the Renaissance period not just in Europe but in the Middle and Far East as well would have trebled the size of the present volume. That being said, it was impossible to ignore the conflicts recorded in the Bible or the great wars fought between the Greek states and Persia.

At one and the same time the third section, *Armies and Soldiers*, includes quotations about the different arms which make up an army and the characteristics of various national armies. In the fourth section, *War and Peace*, are to be found quotations relating to the variegated ingredients of the military life and because the history of warfare is essentially an account of men's efforts to kill each other *Last Post* is a reminder of the more sombre aspects of life — and death — as a soldier.

In making this selection I have re-read many books which have been long familiar to me and in them I found treasures anew. I have also been introduced to many more books previously known to me only by their titles, if at all. One of the pleasures of the compilation has been this introduction to some memorable accounts of war written by some remarkable men. It is extraordinary to consider how, in the midst of war, men have found the language to describe battle and then to commit the words to paper. If we remain fascinated by the resulting literature it is because war is such a cataclysmic event in human experience and one which, thankfully, touches very few men. I am grateful to all those writers who have helped to shed so much light, intelligence and humour on this most significant of subjects, the painful field of battle.

This book is dedicated to my good friend Alan Bold, most unmilitary and pacific of men, and the index was compiled by another good friend, Fiona Ashmore.

Trevor Royle
Edinburgh

To my good friend Alan Bold

I
CAPTAINS AND KINGS

1 AKBAR
Mogul Emperor of India (1542–1605)

1 A monarch should be ever intent on conquest, otherwise his neighbours rise in arms against him.
Akbar, quoted in Abul Fazl, *Aini-i-Akbari* (c. 1590), tr. Jarrett

2 Let not difference of religion interfere with policy, and be not violent in inflicting retribution. Adorn the confidential council with men who know their work. If apologies be made, accept them.
Akbar, letter to his son Murad, governor of Malwa, 1591, quoted in Abul Fazl, *Akbar-nama* (c. 1620), tr. Beveridge

3 Akbar was very great. In any company, at any time, he was a King.
Philip Mason, *The Men Who Ruled India,* vol. I (1953)

4 The very activities which took Akbar away from his books were, in their way, a good preparation for being a soldier king, the only sort of king that it was possible to be in such times.
Bamber Gascoigne, *The Great Moguls* (1971)

5 He was a brilliant general and an outstanding administrator, but his greatness depended on a form of benevolence and sharp common sense.
Geoffrey Moorhouse, *India Britannica* (1983)

2 ALANBROOKE
Alan, Viscount Alanbrooke of Brookeborough, British field marshal (1883–1963)

1 If the stones could talk and could repeat what they have witnessed, and the thoughts they had read on dying men's faces, I wonder if there would ever be any wars.
Field Marshal Viscount Alanbrooke, letter from the Western Front, 1918

2 I have a firm conviction that Right must conquer Wrong.
Field Marshal Viscount Alanbrooke, Diary, 1940

3 Had it not been for the guiding hand of an Almighty Providence, the BEF would never have left the shores of France.
Field Marshal Viscount Alanbrooke, note on the Dunkirk evacuation, 1940, quoted in Bryant, *The Turn of the Tide* (1957)

4 There are times when I wish to God I had not been placed at the helm of a ship that seems to be heading inevitably for the rocks.
Field Marshal Viscount Alanbrooke, letter to his wife, 1942

5 We are certain to have many setbacks to face, many troubles and many shattered hopes, but for all that the horizon is infinitely brighter.
Field Marshal Viscount Alanbrooke, Diary, 1943

6 There is no doubt that from now onwards Russia is all powerful in Europe.
Field Marshal Viscount Alanbrooke, Diary, 1945

7 The patience of a Saint in hardship.
The tenacity of a bulldog in adversity.
The courage of a lion when roused.
The chivalry of a Knight in all his dealings.
Field Marshal Viscount Alanbrooke, 'The ideal soldier', quoted in Fraser, *Alanbrooke* (1982)

8 His work was beyond praise and reward.
Sir James Grigg, quoted in Bryant, *The Turn of the Tide* (1957)

9 The greatest soldier — soldier, sailor or airman — produced by any country taking part in the last war. Those on the inside of affairs would assess his contribution to victory as second only to Churchill.
Field Marshal Viscount Montgomery, quoted in Bryant, *The Turn of the Tide* (1957)

3 ALEXANDER
Harold, Earl Alexander of Tunis, British field marshal (1891–1969)

1 I'm afraid the war will end very soon now, but I suppose all good things come to an end sooner or later, so we mustn't grumble.
Field Marshal Earl Alexander, letter to his Aunt Margaret, 1917

2 I have had a good weathering in this war, and I am going to get the fruits of it if I can. You can bet your life that I shall have a good try.
Field Marshal Earl Alexander, letter to his Aunt Margaret, 1918

3 Surely a soldier on the battlefield, beset by fear and doubt, is far more in need of a guide to action than any games-player at Lord's or Wimbledon?
Field Marshal Earl Alexander, Introduction to Officers' Instruction booklet, I Corps, 1940

4 His Majesty's enemies together with their impedimenta have been completely eliminated from Egypt, Cyrenaica, Libya and Tripolitania. I now await your further instructions.
Field Marshal Earl Alexander, signal to Winston S. Churchill after retaking Tripoli, 1943

5 A commander, if faced by the choice between risking a single soldier's life and destroying a work of art, even a religious symbol like Montecassino, can only make one decision.
Field Marshal Earl Alexander, on his decision to bomb the monastery of Montecassino, 1944, quoted in Nicolson, *Alex: The Life and Times of Field Marshal Earl Alexander of Tunis* (1973)

6 Successful businessmen and soldiers have to depend upon others to achieve results, but the painter is alone with only his colours.
Field Marshal Earl Alexander, letter to Edward Seago, 1969

7 Britain's outstanding soldier in the field of strategy.
General Dwight D. Eisenhower, *Crusade in Europe* (1948)

8 His laziness was a virtue. It meant a capacity to delegate and in wartime it became a tremendous asset, because it meant that he could relax and unhook.
Field Marshal Sir Gerald Templer, quoted in Nicolson, *Alex: The Life of Field Marshal Earl Alexander of Tunis* (1973)

9 He believed that reticence increases a soldier's authority. It was his nature. A gentleman does not utter everything he thinks.
Nigel Nicolson, *Alex: The Life of Field Marshal Earl Alexander of Tunis* (1973)

10 He was the last of our gentlemanly generals.
Jan Morris, in *The Times*, 1983

4 ALEXANDER THE GREAT
King of Greece (356–323 BC)

1 What of the two men in command? You have Alexander, they — Darius!
Alexander the Great, address to his army before the Battle of Issus, 333 BC, quoted in Arrian, *The Campaigns of Alexander the Great*, II, 7 (*c.* AD 150), trs. de Selincourt

2 I will not steal a victory.
   Alexander the Great, refusing to fall on the army of King Darius before the Battle of Arbela, 331 BC, quoted in Plutarch, *Lives*

3 Go home and tell them that you left Alexander to conquer the world alone.
   Alexander the Great, address to his disaffected soldiers in India, 326 BC, quoted in Plutarch, *Lives*

4 Stand firm; for well you know that hardship and danger are the price of glory, and that sweet is the savour of a life of courage and of deathless renown beyond the grave.
   Alexander the Great, address to his men after the mutiny of the Macedonian army, 324 BC, quoted in Arrian, *The Campaigns of Alexander the Great*, V, 26 (*c.* AD 150), trs. de Selincourt

5 My son, seek thee out a kingdom equal to thyself; Macedonia has not room for thee.
   Philip of Macedon, quoted in Plutarch, *Lives*.

6 Alexander made no fresh appointment to the command of the Companion cavalry; he wished Hephaestion's name to be preserved always in connection with it, so Hephaestion's Regiment it continued to be called and Hephaestion's image continued to be carried before it.
   Arrian, *The Campaigns of Alexander the Great*, VII, 15 (*c.* AD 150), trs. de Selincourt

7 In the course of many campaigns he captured fortified towns, slaughtered kings, traversed the earth to its remotest bounds, and plundered innumerable nations.
   1 Maccabees, 1:2

8 When in the world I lived I was the world's commander
   By east, west, north and south I spread my conquering might
   My scutcheon plain declares that I am Alexander.
   William Shakespeare, *Love's Labour Lost*, V, ii

9 Alexander at the head of the world never tasted the true pleasure that boys of his own age have enjoyed at the head of the school.
   Horace Walpole, letter to Edward Montagu, 1736

---

5 ALFRED THE GREAT
King of Wessex (849–99)

---

1 A man may fight on behalf of his born kinsman, if anyone attacks him wrongfully, unless it is against his lord: that we do not allow.
   King Alfred the Great, Law Code, *c.* 880

2 This I can now truly say, that so long as I have lived, I have striven to live worthily, and after my death to leave my memory to my descendants in good works.
   King Alfred the Great, translation of Boethius, *De Consolatione Philosophiae* (*c.* 897)

3 Truth-teller was our England's Alfred named.
   Alfred, Lord Tennyson, 'Ode on the Death of the Duke of Wellington' (1852)

4 There was not any English Armour left,
   Nor any English thing,
   When Alfred came to Athelny
   To be an English King.
   G.K. Chesterton, *The Ballad of the White Horse*, I (1911)

5 A mythical monarch with many of the gifts of Napoleon, and most of the qualities of Abraham Lincoln.
   Philip Guedalla, *Men of Affairs, Men of War, Still Life* (1927)

---

6 ALLENBY
Edmund Henry Hynman, Viscount Allenby of Megiddo, British field marshal (1861–1936)

---

1 One is always liable to be smashed by superior force, but one should never be caught unprepared to do one's best.
   Field Marshal Viscount Allenby, letter to his wife from the Boer War, 1900

2 Think to a finish!
   Field Marshal Viscount Allenby, advice to officers of 5th Lancers, on taking command, 1902

3 I have now the biggest command of cavalry that anyone has had in the history of the army, and can see that I have the best trained and most efficient officers and men that have ever taken the field in European war. I have also a first class staff. So I have no excuse if I do badly.
Field Marshal Viscount Allenby, letter to his wife, 1914

4 In pursuit you must always stretch possibilities to the limit. Troops having beaten the enemy will want to rest. They must be given as objectives, not those that you think they will reach, but the farthest they could possibly reach.
Field Marshal Viscount Allenby, Order to 21st Corps, Philistia, 1917

5 I have never been in a *difficult* position in my life. I have sometimes been in an *impossible* one, and then I got out of it as quickly as I could.
Field Marshal Viscount Allenby, address to his staff, Cairo, 1919

6 Misunderstandings and petty quarrels between individuals often occur in even the happiest families, but they are composed amicably without resort to knife or pistol. So it should be in the case of bickering between nations.
Field Marshal Viscount Allenby, Rectorial Address, Edinburgh University, 1936

7 Once you have taken a decision, never look back on it.
Field Marshal Viscount Allenby, favourite maxim, quoted in Wavell, *Allenby, Soldier and Statesman* (1946)

8 Allenby was the image we worshipped.
T.E. Lawrence, *The Seven Pillars of Wisdom* (1935)

9 The last of the Paladins.
Sir Ronald Storrs, *Orientations* (1937)

10 The British Army has had few leaders with better mental or physical equipment for the tough test of war, less likely to lose heart in the darkest hour, or more remorseless in pressing home an advantage and completing a victory; certainly none with a greater sense of loyalty and duty or more of the truth and straightforwardness that mark a great and generous nature.
Field Marshal Earl Wavell, *Allenby, Soldier and Statesman* (1946)

7 ALVA
Fernando Alvarez de Toledo, Duke of Alva, Spanish soldier (1503–83)

1 I have tamed men of iron and why then shall I not be able to tame these men of butter?
Duke of Alva, reply to King Philip II of Spain on being appointed Governor-General of the Netherlands, 1567

2 Putting in new men one by one is like throwing a bottle of good wine into a vat of vinegar. The wine will instantly turn to vinegar too.
Duke of Alva, letter to King Philip II of Spain explaining his reasons not to appoint Dutch officials, 1567, quoted in *Papeles de la secretario de estado* (1573)

3 A great deal remains to be done first. The towns must be punished for their rebelliousness with the loss of privileges; a goodly sum must be squeezed out of private persons; a permanent tax obtained from the states of the country. It would therefore be unsuitable to proclaim a pardon at this juncture.
Duke of Alva, letter to King Philip II of Spain outlining his policy for the Netherlands after the First Dutch Revolt, 1568, quoted in Alba, *Epistilario*, III (1952)

4 These troubles must be ended by force of arms without any use of pardon, mildness, negotiations or talks until everything has been flattened. That will be the right time for negotiation.
Duke of Alva, letter to Don Luis de Requesens, his successor as Governor-General of the Netherlands, during the Second Dutch Revolt, 1573, quoted in Abal, *Epistilario*, III (1952)

5 Hellish father who in Brussels doth dwell Cursed be thy name in heaven and in hell; Thy kingdom, which has lasted too long, be gone,
Thy will in heaven and earth be not done.
Anon, 'Paternoster of Ghent', composed 'In honour of the Duke of Alva', 1572, quoted in Motley, *Rise of the Dutch Republic* (1882)

8 ATAHUALPA
Inca Emperor (1502–33)

1 My God still lives.
Atahualpa, before the massacre of the Incas at Cajmarca, 1532

2 Atahualpa was loved by the old captains of his father and the soldiers, because he went to the wars with them as a child.
Pedro de Cieza de Leon, *The Incas* (1554), trs. de Onis

3 He did not know the character of the Spaniards.
William Hickling Prescott, *The Conquest of Peru* (1847)

9 ATTILA THE HUN
Barbarian leader (406–53)

1 Descendant of the Great Nimrod. Nurtured in Engaddi. By the Grace of God, King of the Huns, the Goths, the Danes and the Medes. The Dread of the World.
Attila, designation of himself, quoted in Creasey, *Fifteen Decisive Battles of the Western World* (1908)

2 Grass never grows again where my horse has once trodden.
Attila, attr.

3 Attila ground almost the whole of Europe into the dust.
Count Marcellinus, quoted in Thompson, *A History of Attila and the Huns* (1948)

4 Looke, Attila, the gret conquerour . . .
Deyde in his sleep, with shame and
          dishonour,
Bledynge ay at his nose in dronkenesse.
Geoffrey Chaucer, *The Canterbury Tales*, 'The Pardoner's Tale' (*c.* 1327)

5 He delighted in war; but, after he had ascended the throne in a mature age, his head rather than his hand achieved the conquest of the North; and the fame of an adventurous soldier was usefully exchanged for that of a prudent and successful general.
Edward Gibbon, *The History of the Decline and Fall of the Roman Empire*, III (1776)

10 AUCHINLECK
Sir Claude Auchinleck, British field marshal (1884–1981)

1 We shall not win this war as long as we cling to worn-out shibboleths and snobberies.
Field Marshal Sir Claude Auchinleck, letter to General Sir Robert Haining, 1940

2 It is not sound to take an unreasonable risk.
Field Marshal Sir Claude Auchinleck, letter to Winston S. Churchill, 1942

3 No Indian officer must be regarded as suspect and disloyal merely because he is what is called a 'Nationalist', or in other words, a good Indian.
Field Marshal Sir Claude Auchinleck, memorandum to South-East Asia Command, 1946

4 I doubt if runs before breakfast really produce battle winners of necessity.
Field Marshal Sir Claude Auchinleck, quoted in Connell, *Auchinleck: A Critical Biography* (1959)

5 I don't like being separated in comparative luxury from troops who are living hard.
Field Marshal Sir Claude Auchinleck, interview with David Dimbleby, 1976

6 I think soldiering was in my blood and I really cared for nothing else.
Field Marshal Sir Claude Auchinleck, interview with David Dimbleby, 1976

7 (Of his leadership in North Africa) History will prove me right.
Field Marshal Sir Claude Auchinleck, quoted in Warner, *Auchinleck: The Lonely Soldier* (1981)

8 Of General Auchinleck I will only say that he is an officer of the greatest distinction, and a character of singular elevation.
Winston S. Churchill, speech to House of Commons, 1942

9 The soldiers loved him because he was such an honest-to-God soldier.
Colonel Jeff Alexander, quoted in Warner, *Auchinleck: The Lonely Soldier* (1981)

10 The only battle he lost was for his own fulfilment.
Philip Warner, *Auchinleck: The Lonely Soldier* (1981)

11 BABUR
Mogul Emperor of India (1483–1530)

1 In war and affairs of state many things seem to be just and reasonable at first sight; yet nothing of the kind ought to be finally decided without being pondered in a hundred different lights.
Babur, *Babur-nama* (1526–30), trs. Beveridge

2 (Of the loss of Samarkand, 1498) It came very hard on me, I could not help crying a great deal.
Babur, *Babur-nama* (1526–30), trs. Beveridge

3 Most men, however brave, have some anxiety or fear in them.
Babur, *Babur-nama* (1526–30), trs. Beveridge

4 The treasures of five kings fell into his hands; he gave everything away.
Gulbadan, *Humayun-nama* (1587), trs. Beveridge

5 In his rule, with its ruthless force during conquests and unexpected tolerance afterwards, the tradition of the great Mongol Khans can be traced.
Harold Lamb, *Babur the Tiger* (1961)

6 This first Mughal took his new domain by the sword (from occupying Afghans) and held it by military might (against resisting Indian Rajputs).
Geoffrey Moorhouse, *India Britannica* (1983)

12 BADEN-POWELL
Robert Stephenson Smith, Baron Baden-Powell of Gilwell, British general (1857–1941)

1 All well, four hours' bombardment. One dog killed.
Lord Baden-Powell, despatch at the start of the Siege of Mafeking, 1899

2 Beneath the British flag all men are free.
Lord Baden-Powell, *Scouting for Boys* (1908)

3 Country first, self second.
Lord Baden-Powell, *Scouting for Boys* (1908)

4 Make yourselves good scouts and good rifle shots in order to protect the women and children of your country if it should ever become necessary.
Lord Baden-Powell, *Scouting for Boys* (1908)

5 Our effort is not so much to discipline the boys as to teach them to discipline themselves.
Lord Baden-Powell, *Scouting for Boys* (1908)

6 We have all got to die some day; a few years more or less of our lives don't much matter if by dying a year or two sooner than we should otherwise do from disease we can help to save the flag of our country from going under.
Lord Baden-Powell, *Scouting for Boys* (1908)

7 Personally, I should like to see all Boy Scouts drilled. I look upon the Movement as a further saving of the situation for the nation in the future, and that it will pave the way directly for its national service.
Lord Baden-Powell, letter to Lord Meath, 1910

8 Our business is not merely to keep up smart 'show' troops, but to pass as many boys through our character factory as we can.
Lord Baden-Powell, *Headquarters Gazette* (1911)

9 We want to make our nation entirely one of gentlemen, men who have a strong sense of honour, of chivalry towards others, of playing the game bravely and unselfishly for their side, and to play it with a sense of fair play and happiness for all.
Lord Baden-Powell, *Headquarters Gazette* (1914)

10 B-P would have made an ideal headmaster in a Victorian adventure story. A ripper when the going was good, but an alarming man to have as your enemy.
Thomas Pakenham, *The Boer War* (1979)

BARBAROSSA
See 53. FREDERICK I

13 BERNADOTTE
Jean Baptiste Jules, Marshal of France (1763–1844)

1 The fame of battles won belongs more to the brave warriors, more to the generals who encouraged them, than to the Minister.
Marshal Jean Baptiste Bernadotte, letter to the Directory after being sacked as Minister of War, 1799

2 A soldier's happiness lies in discharging his duty.
Marshal Jean Baptiste Bernadotte, address to the Saxon Army on their disbandment, 1808, quoted in Barton, *Bernadotte* (1914–25)

3 One may for a time suppress the feeling for independence, but one can never quite remove it from the hearts of nations.
Marshal Jean Baptiste Bernadotte, letter to Napoleon withdrawing Sweden's support from France, 1813

14 BISMARCK
Count Otto von, Chancellor of Germany (1815–98)

1 Anyone who has ever looked into the glazed eyes of a soldier dying on the battlefield will think hard before starting a war.
Count Otto von Bismarck, speech in Berlin, 1867

2 This policy cannot succeed through speeches and shooting competitions and songs; it can succeed only through blood and iron.
Count Otto von Bismarck, speech in Prussian Lower House, 1886

3 Dropping the pilot.
Sir John Tenniel, caption of a cartoon in *Punch* referring to Bismarck's departure from office, 1890

4 He had been as ruthless and unscrupulous as any other politician. What had distinguished him was his moderation.
A.J.P. Taylor, *The Course of German History* (1950)

5 He lied with consistency and with enjoyment, although unlike Lenin, he did not actually prefer lying to telling the truth.
Edward Crankshaw, *Bismarck* (1981)

BLACK PRINCE
See 45. EDWARD

15 BLÜCHER
Gebhard Liebrecht von, Prinz von Wahlstadt, Prussian field marshal (1742–1819)

1 What a place to plunder!
Marshal Blücher, on visiting London, 1814

2 *Mein lieber Kamerad! Quelle affaire!*
Marshal Blücher, greeting the Duke of Wellington after the Battle of Waterloo, 1815

3 What I promised I have performed. On the 16th I was forced to fall back a short distance; the 18th — in conjunction with my good friend Wellington — completed Napoleon's ruin.
Marshal Blücher, letter to his wife, 1815

4 Blücher and his Prussians marched steadily and uneventfully upon Paris.
Winston S. Churchill, *A History of the English Speaking Peoples*, vol. III (1957)

16 BOADICEA
Queen Boudicca of the Iceni (d. AD 61)

1 We British are used to women commanders in war. I am not fighting for my kingdom and wealth now. I am fighting as an ordinary person for my lost freedom, my bruised body and my outraged daughters.
Queen Boudicca, address to her army before the Icenian revolt, AD 61, quoted in Tacitus, *Annals*

2 Kingdom and household alike were plundered like prizes of war, the one by Roman officers, the other by Roman slaves. As a beginning, the widow Boadicea was flogged and her daughters raped.
Tacitus, *Annals*

3 Princess! if our aged eyes
Weep upon thy matchless wrongs,

'Tis because resentment ties
All the terrors of our tongues.
William Cowper, the Seer's advice to
Boadicea, *Boadicea: An Ode* (1780)

4 Up my Britons, in my chariot, on my
chargers, trample them under us!
Lord Tennyson, Boadicea's call to arms,
Boadicea (1864)

5 The rising of Boadicea is the exception that
proves the rule of the easy submission of
East and South to the Roman influence.
G.M. Trevelyan, *History of England* (1926)

17 BOLIVAR
Simon, South American soldier
(1783–1830)

1 Our hatred knowns no bounds, and the war
shall be to the death.
Simon Bolivar, proclamation to his army
before the drive to Caracas, 1812, quoted in
Ybarra, *Bolivar* (1929)

2 A people that loves freedom will in the end
be free.
Simon Bolivar, *Letter from Jamaica* (1815)

3 War lives on despotism and is not waged
with God's love.
Simon Bolivar, address to his army at
Trujillo, 1823, quoted in *Memorias del
General Daniel Florencio O'Leary*, trs.
McNerney (1970)

BONNIE PRINCE CHARLIE
See 29. CHARLES EDWARD STUART

18 BRADLEY
Omar Nelson, American general
(1893–1981)

1 Unless each officer and non-commissioned
officer has capabilities greatly in excess of
the responsibility he holds, he is basically
an unprofitable part of a military machine.
General Omar N. Bradley, *Military
Review* (1950)

2 (Of the Korean War) The wrong war, at the
wrong place, at the wrong time and with
the wrong enemy.
General Omar N. Bradley, speech to the
Senate, 1951

3 Fairness, diligence, sound preparation,
professional skill and loyalty are the marks
of American military leadership.
General Omar N. Bradley, *Army
Information Digest* (1953)

4 Heritage is not freedom alone, but rather
freedom with responsibility. Responsibility
of the individual to the nation.
General Omar N. Bradley, address on
Veterans Day, Los Angeles, 1969

5 Freedom — no word was ever spoken that
has held out greater hope, demanded
greater sacrifice, needed more to be
nurtured, blessed more than the giver,
damned more than its discharge, or come
closer to being God's will on earth. And I
think that's worth fighting for.
General Omar N. Bradley, address to the
National Press Club, Washington, 1970

6 Leadership in a democratic army means
firmness, not harshness; understanding,
not weakness; justice not license;
understanding, not intolerance; generosity,
not selfishness; pride, not egotism.
General Omar N. Bradley, quoted by
General Westmoreland in a speech at
Abilene, Texas, 1971

7 Bradley was an able tactician, but he was
less competent in the realm of strategy.
Chester Wilmot, *The Struggle for Europe*
(1952)

19 BRIAN BORU
High King of Ireland (d. 1014)

1 Flight becomes me not and I myself know
that I shall not go from here alive, and what
should it profit me though I did?
King Brian Boru, before the Battle of
Clontarf, 1014, quoted in 'The Wars of the
Gael with the Gaill', *Book of Leinster*
(1150), trs. Todd

2 Oh where, Kincora, is Brian the great,
And where is the beauty that once was
thine,
Oh where are the princes and nobles that
sate
To feast in thy halls and drink the red wine,
Where, O Kincora?
Mac Liag, 'Lament for Brian Boru', 1014,
quoted in Hyde, *The Story of Early Gaelic
Literature* (1895)

3   Swordblades rang on Ireland's coast,
    Metal yelled as shield it sought,
    Spear-points in the well-armed host.
    I heard sword-blades many more;
    Sigurd fell in battle's blast,
    From his wounds there sprang hot gore.
    Brian fell, but won at last.
    Njal's Saga, on the death of Brian Boru at
    the Battle of Clontarf, 1014

20 BURGOYNE
John, British general (1722–92)

1   After a fatal procrastination, not only of
    vigorous measures, but of preparations for
    such, we took a step as decisive as the
    passage of the Rubicon, and now find
    ourselves plunged at once in a most serious
    war without a single requisition, gunpowder
    excepted, for carrying it on.
    Lt-Gen John Burgoyne, letter to Lord
    Germain after the Battle of Lexington, 1775

2   This Army must not retreat.
    Lt-Gen John Burgoyne, Order to the
    Anglo-Hessian Army before the Battle of
    Ticonderoga, 1777, quoted in de
    Fonblanque, *Political and Military Episodes*
    (1876)

3   The English never lose ground.
    Lt-Gen John Burgoyne, before the defeat of
    his army at the Battle of Saratoga, 1777,
    quoted in von Riedesel, *Letters and Journals*
    (1867)

4   Burgoyne's surrender at Saratoga made him
    that occasionally necessary part of our
    British system, a scapegoat.
    George Bernard Shaw, *The Devil's Disciple*
    (1901)

21 CAMBRIDGE
**George William Frederick Charles, Duke
of Cambridge, British field marshal
(1819–1904)**

1   War is indeed a fearful thing and the more I
    see it the more dreadful it appears.
    Duke of Cambridge, letter to his wife from
    the Crimea, 1854

2   An army never can be commanded or
    controlled by civilians.
    Duke of Cambridge, letter to Sir Colin
    Campbell, 1858

3   Why should we want to know anything
    about foreign cavalry? We have better
    cavalry of our own. I fear, gentlemen, that
    the Army is in danger of becoming a mere
    debating society.
    Duke of Cambridge, introducing a lecture
    on foreign cavalry at Aldershot, 1881,
    quoted in Robertson, *From Private to Field
    Marshal* (1921)

4   Brains, I don't believe in brains. You
    haven't any, I know, Sir.
    Duke of Cambridge, remark to a divisional
    General during an inspection, 1890,
    quoted in Callwell, *Stray Recollections*, I
    (1923)

5   I have been told that I am now a fifth wheel
    to the coach, and that it is time that I made
    way for a man who is younger and more
    closely in touch with the military
    requirements of the present day.
    Duke of Cambridge, speech to the Staff
    College, Camberley, announcing his
    resignation as Commander-in-Chief, 1892

6   Educated to believe in the Army as he
    found it, because it had been made by the
    Great Duke of Wellington, he honestly and
    firmly believed that what had been created
    by such a master of war must be the best
    for all time.
    Field Marshal Viscount Wolseley, *The
    Story of a Soldier's Life* (1903)

22 CAMPBELL
**Colin, Baron Clyde of Clydesdale, British
field marshal (1792–1863)**

1   By means of patience, common-sense and
    time, impossibility becomes possible.
    Field Marshal Sir Colin Campbell,
    inscription in his memorandum book, 1832

2   No soldier must go carrying off wounded
    men. If any soldier does such a thing his
    name shall be stuck up in this parish church.
    Field Marshal Sir Colin Campbell, Order to
    the Highland Brigade before the Battle of
    Alma, 1854

3 We'll hae nane but Hielan bonnets here.
Field Marshal Sir Colin Campbell, Order to
the Highland Brigade before the Battle of
Alma, 1854

4 Remember, there is no retreat from here.
You must die where you stand.
Field Marshal Sir Colin Campbell, Order to
the Highland Brigade before the Battle of
Balaclava, 1854

23 CARDIGAN
**Thomas James Brudenell, Earl of
Cardigan, British general (1797–1868)**

1 Damn those Heavies! They have the laugh
of us this day!
Lt-Gen Earl of Cardigan, before the Battle
of Balaclava, 1854

2 Well, here goes the last of the Brudenells.
Lt-Gen Earl of Cardigan, before leading the
Charge of the Light Brigade at the Battle of
Balaclava, 1854

3 Men, it is a mad-brained trick, but it is no
fault of mine.
Lt-Gen Earl of Cardigan, address to his men
after the Charge of the Light Brigade at the
Battle of Balaclava, 1854 .

4 In Lord Cardigan there was such an absence
of guile that exactly as he was so he showed
himself to the world. Of all false pretences
contrived for the purpose of feigning an
interest in others he was as innocent as a
horse.
A.W. Kinglake, *The Invasion of the Crimea,*
vol. IV (1863–87)

24 CARNOT
**Lazare Nicolas Marguerite, French general
and minister of war (1753–1823)**

1 I am a soldier, I speak little, and I don't
belong to any party.
Lazare Carnot, address to Jacobins Club,
1791

2 The situation is summed up in four words:
'We have not anything.'
Lazare Carnot, speech to the Legislative
Assembly on the condition of the French
Army, 1792

3 Remember, soldiers, that first and foremost
you are citizens. Let us not become a greater

scourge to our country than the enemy
themselves.
Lazare Carnot, address to Armée du Nord,
1792

4 All Frenchmen are in permanent requisition
for army service. The young men will go to
fight; the married men will forge arms and
carry supplies; the women will make tents
and uniforms and will serve in the hospitals;
the children will shred old clothes; the old
men will be taken to the public squares to
excite the courage of the combatants, the
hatred of royalty and the unity of the
Republic.
Lazare Carnot, Mobilization Decree, Paris,
1793

5 It is necessary to learn the art of attacking
the enemy always where he is weak, using
such a superiority of force that victory can
never be in doubt.
Lazare Carnot, letter to General Pichegru,
1793

6 *L'Organisateur de la Victoire.* The organiser
of victory.
Deputy Lanjuinais, speech in defence of
Carnot in the Legislative Assembly, 1795

7 Cold mathematical head and silent
stubbornness of will: iron Carnot,
far-planning, imperturbable, unconquerable,
who in the hour of need shall not be found
wanting.
Thomas Carlyle, *The French Revolution*
(1837)

25 CHARLES I
King of England (1600–49)

1 Never make a defence of apology before you
be accused.
King Charles I, letter to Sir Thomas
Wentworth, 1636, quoted in Knowler, *The
Earl of Strafford's Letters and Dispatches*
(1739)

2 I am going to fight for my crown and my
dignity.
King Charles I, before the Battle of Edgehill,
1642, quoted in Gardiner, *History of
England* (1893)

3 I have set my rest on the Justice of my cause,
being resolved that no Extremity or

Misfortune shall me yield. For either I will be a Glorious King or a patient Martyr.
King Charles I, letter to the Duke of Hamilton after the Battle of Turnham Green, 1642, quoted in Rushworth, *Historical Collections of Private Passages of State*, V (1680–1701)

4 As a Christian, I must tell you that God will not suffer Rebels and Traytors to prosper nor his Cause to be overthrown.
King Charles I, letter to Prince Rupert, 1645, quoted in Clarendon, *History of the Rebellion* (1704)

5 A mild and gracious prince who knew not how to be, or how to be made, great.
Archbishop William Laud, quoted in Heylin, *Cyprianus Angelicus* (1688)

6 He was very fearless in his person, but not enterprising, and had an excellent understanding, but was not confident enough of it; which made him often times change his opinion for a worse, and follow the advice of a man that did not judge so well as himself.
Earl of Clarendon, *History of the Rebellion* (1704)

26 CHARLES II
King of England (1630–85)

1 We have so deep a sense of the present miseries and calamities of this kingdom, that there is nothing that we more earnestly pray to almighty God than that He would be pleased to restore unto it a happy peace.
King Charles II, letter to Sir Thomas Fairfax, 1645

2 It is upon the navy, under the Providence of God, that the safety, honour and welfare of this realm do chiefly attend.
King Charles II, *Articles of War* (1652)

3 Men ordinarily become more timid as they grow old; as for me, I shall be, on the contrary, bolder and firmer and I will not stain my life and reputation in the little time that perhaps remains for me to live.
King Charles II, before the opening of Parliament in Oxford, 1681

4 He seemed to be chiefly desirous of 'Peace and Quiet for his own time.'
Sir John Reresby, 1676, quoted in *Memoirs* (1735)

5 King Charles the Second in the Oak near Boscobel makes as Historical Figure as in any part of his Reign.
John Oldmixon, *History of England* (1731)

27 CHARLES V
Holy Roman Emperor (1500–48)

1 O God! I promise Thee that this war has not been begun by me, and that the King of France seeks to make me greater than I am.
Emperor Charles V, on hearing that French soldiers had crossed the Pyrenees to invade Spain, 1521

2 You will be king if you rule justly; otherwise you will no longer be king.
Emperor Charles V, maxim adopted at the time of his coronation at Bologna, 1530

3 I make war on the living, not on the dead.
Emperor Charles V, after being advised to hang Martin Luther's corpse on a gallows, 1546

4 I came; I saw; and God conquered.
Emperor Charles V, after the Battle of Mühlberg, 1547, quoted in Avila, *Commentarium de bello Germanico a Carolo Quinto Caesare Maxima Gesto* (1548)

5 Name me an emperor who was ever struck by a cannon-ball.
Emperor Charles V, remark after the Battle of Mühlberg, 1547

6 We human beings act according to our powers, our strength, our spirit, and God awards victory or permits defeat.
Emperor Charles V, speech of abdication, Brussels, 1555

7 My house will survive because it is one with the spirit of the universe of which it is the centre, whence it can patiently await the return of erring spirits.
Emperor Charles V, attr., *c.* 1556, quoted in Grillparzer, *König Ottokar's Glück und Ende* (1825)

8 Depend on none but yourself.
Emperor Charles V, maxim for his son King Philip II of Spain, *c.* 1558

28 CHARLES XII
King of Sweden (1682–1718)

1 That shall be my music in the future!
King Charles XII, on first hearing the whistling of bullets at Copenhagen, 1700

2 So long as a trooper is a good soldier it can be of no consequence whether or no he be a man of family.
King Charles XII, refusing a commission to a nobleman after the Siege of Grodno, 1706

3 A man may cease to be lucky, for that is beyond his control; but he should not cease to be honest.
King Charles XII, to Axel von Löwen, 1714

4 I wish I could serve in some campaign under so great a commander that I might learn what yet I want to know in the art of war.
Duke of Marlborough, letter to King Charles XII, 1707

5 The only person in history who was free from all human weakness.
Voltaire, *Life of Charles XII* (1731)

6 Heroes are much the same, the point's agreed
From Macedonia's madman to the Swede.
Alexander Pope, *An Essay on Man*, IV, 219 (1732–4)

7 He left the name at which the world grew pale,
To point a moral or adorn a tale.
Samuel Johnson, *The Vanity of Human Wishes* (1749)

29 CHARLES EDWARD STUART
The Young Pretender ('Bonnie Prince Charlie') (1720–88)

1 I am come home, sir, and I will entertain no notion at all of returning to that place whence I came, for I am persuaded my faithful Highlanders will stand by me.
Prince Charles Edward Stuart, on landing at Moidart, 1745

2 I may be overcome by my enemies, but I will not dishonour myself; if I die, it shall be with my sword in my hand, fighting for the liberty of those who fight against me.
Prince Charles Edward Stuart, letter to his father from Perth, Scotland, 1745

3 If I had obtained this victory over foreigners my joy would have been complete; but as it is over Englishmen, it has thrown a damp upon me that I little imagined.
Prince Charles Edward Stuart, letter to his father after the Battle of Prestonpans, 1745

4 Had Prince Charles Edward slept during the whole of the expedition and allowed Lord George Murray to act for him according to his own judgement, there is every reason for supposing he would have found the crown of Great Britain on his head when he awoke.
Chevalier James Johnstone, *Memoirs* (1820)

5 Charlie is my darling, my darling, my darling,
Charlie is my darling, the young Chevalier.
Lady Nairne, 'Charlie is my Darling', *Works* (1869)

6 Biography scarcely records a dawn more brilliant, a sunset more clouded.
A.C. Ewald, *Life and Times of Prince Charles Stuart* (1875)

30 CHIANG KAI-SHEK
Chinese general (1887–1975)

1 We are working for a revolution. If we do not start it by improving the life of the soldiers, all slogans of reforming and improving society are but empty words.
Generalissimo Chiang Kai-Shek, letter to Chou En-lai, 1925

2 We shall not lightly talk about sacrifice until we are driven to the last extremity which makes sacrifice inevitable.
Generalissimo Chiang Kai-Shek, speech to Fifth Congress of the Kuomintang, 1935

3 The enemy never realises that China's territory is not conquerable. She is indestructible. As long as there is one spot in China free from enemy encroachment, the National Government will remain supreme.
Generalissimo Chiang Kai-Shek, interview in the *New York Times*, 1937

4 To our new common battle we offer all we are and all we have, to stand with you until the Pacific and the world are freed from the curse of brute force and endless perfidy.
Generalissimo Chiang Kai-Shek, telegram to President Roosevelt, following China's declaration of war on Japan and Germany, 1941

5 The Japanese are a disease of the skin. The Communists are a disease of the heart.
Generalissimo Chiang Kai-Shek, 1942, quoted in Crozier, *The Man who Lost China* (1976)

31 CHURCHILL
Sir Winston Spencer, British prime minister (1874–1965)

1 I would say to the House, as I said to those who have joined this Government, 'I have nothing to offer but blood, toil, tears and sweat.'
Winston S. Churchill, speech in House of Commons, 1940

2 Let us therefore brace ourselves to our duties and so bear ourselves that, if the British Empire and its Commonwealth last for a thousand years, men will still say: 'This was their finest hour.'
Winston S. Churchill, speech in House of Commons, 1940

3 (Of the Battle of Britain) Never in the field of human conflict was so much owed by so many to so few.
Winston S. Churchill, speech in House of Commons, 1940

4 Victory at all cost, victory in spite of all terror, victory however long and hard the road may be; for without victory there is no survival.
Winston S. Churchill, speech in House of Commons, 1940

5 We shall defend our island, whatever the cost may be, we shall fight on the beaches, we shall fight on the landing grounds, we shall fight in the fields and in the streets, we shall fight in the hills; we shall never surrender.
Winston S. Churchill, speech in House of Commons, 1940

6 Give us the tools, and we will finish the job.
Winston S. Churchill, broadcast addressed to President Roosevelt, 1941

7 (Of the victory in Egypt) This is not the end. It is not even the beginning of the end. But it is, perhaps, the end of the beginning.
Winston S. Churchill, speech at the Mansion House, 1941

8 (Of the French government) When I warned them that Britain would fight on alone whatever they did, their Generals told their Prime Minister and his divided Cabinet: 'In three weeks England will have her neck wrung like a chicken.' Some chicken! Some neck!
Winston S. Churchill, speech to the Canadian Parliament 1941

9 I have not become the King's First Minister in order to preside over the liquidation of the British Empire.
Winston S. Churchill, speech at the Mansion House, 1942

10 'Not in vain' may be the pride of those who survived and the epitaph of those who fell.
Winston S. Churchill, speech in House of Commons, 1944

11 An iron curtain has descended across the Continent.
Winston S. Churchill, speech at Westminster College, Fulton, USA, 1946

12 In War: Resolution. In Defeat: Defiance. In Victory: Magnanimity. In Peace: Goodwill.
Winston S. Churchill, 'Moral of the Work', *The Second World War* (1948)

13 No one can guarantee success in war, but only deserve it.
Winston S. Churchill, *The Second World War*, vol. II (1949)

14 To jaw-jaw is better than to war-war.
Winston S. Churchill, speech in Washington, 1954

15 He has spoiled himself by reading about Napoleon.
David Lloyd George, quoted in Frances Stevenson's diary, 1917

16 Churchill has the habit of breaking the rungs of any ladder he puts his foot to.
Lord Beaverbrook, letter to Arthur Brisbane, 1932

17 Winston's back.
Admiralty message to the Royal Navy on
the appointment of Churchill as First Sea
Lord, 1939

18 His passion for the combative renders him
insensitive to the gentle gradations of the
human mind.
Harold Nicolson, Diary, 1945

19 Churchill on top of the wave has in him the
stuff of which tyrants are made.
Lord Beaverbrook, *Politicians and the War*
(1959)

32 CLARK
**Mark Wayne, American general
(1887–1975)**

1 We're spitting into the lion's mouth and we
know it.
General Mark Clark, press briefing before
the Salerno landings, 1943

2 The Americans have carried the ball every
inch of the way in this battle, dragging the
British troops on their flank.
General Mark Clark, Diary, before the
Battle for Bologna, 1944

3 (Of the Italian campaign) We are caught in
the British Empire machine.
General Mark Clark, Diary, 1944

4 We must build a national military team that
will make it unavoidably clear that anybody
who endangers our way of life will risk
destroying himself.
General Mark Clark, letter to his wife,
1950

5 The battle of Cassino was the most
gruelling, the most harrowing, and in one
aspect perhaps the most tragic of any phase
of the war in Italy.
General Mark Clark, *Calculated Risk*
(1951)

6 I know from intimate personal experience
that evil forces are loose in the world, and
that their sinister aim is to destroy our free
way of life — by subversion if possible, by
overt hostility if necessary.
General Mark Clark, address to Cold War
Seminar, Charleston, South Carolina, 1961

7 The more stars you have, the higher you

climb the flagpole, the more of your ass is
exposed.
General Mark Clark, quoted in Blumenson,
Mark Clark (1984)

33 CLAUSEWITZ
**Karl Maria von, Prussian philosopher of
war (1780–1831)**

1 On no account should we overlook the
moral effect of a rapid, running assault. It
hardens the advancing soldier against
danger, while the stationary soldier loses his
presence of mind.
Karl von Clausewitz, *Principles of War*
(1812)

2 Never forget that no military leader has ever
become great without audacity.
Karl von Clausewitz, *Principles of War*
(1812)

3 A fundamental principle is never to remain
completely passive, but to attack the enemy
frontally and from the flanks, even while he
is attacking us.
Karl von Clausewitz, *On War* (1832)

4 It is even better to act quickly and err than
to hesitate until the time of action is past.
Karl von Clausewitz, *On War* (1832)

5 It is politics which beget war. Politics
represents the intelligence, war merely its
instrument, not the other way round. The
only reasonable course in war is to
subordinate the military viewpoint to the
political.
Karl von Clausewitz, *On War* (1832)

6 No one starts a war — or rather, no one in
his senses should do so — without being
clear in his mind what he intends to achieve
by that war and how he intends to conduct
it.
Karl von Clausewitz, *On War* (1832)

7 War belongs not to the Arts and Sciences,
but to the provenance of social life.
Karl von Clausewitz, *On War* (1832)

8 War is not merely a political act, but also a
political instrument, a continuation of
political relations, a carrying out of the same
by other means.
Karl von Clausewitz, *On War* (1832)

9 This fellow has a common-sense that
   borders on wittiness.
   Karl Marx, letter to Friedrich Engels, 1858

10 A doctrine which began by defining war as
   only a continuation of state policy by other
   means led to the contradictory end of
   making policy the slave of strategy.
   Captain Sir Basil Liddell Hart, *Strategy*
   (1954)

CLAVERHOUSE
See 63. GRAHAM

34 CLIVE
   **Robert, Baron Clive, British soldier and
   administrator (1725–74)**

1 It appears I am destined for something; I
   will live.
   Robert, Lord Clive, after his unsuccessful
   suicide attempt, 1744

2 This expedition, if attended with success,
   may enable me to do great things. It is by far
   the grandest of my undertakings. I go with
   great forces and great authority.
   Robert, Lord Clive, letter to his father
   before the capture of Calcutta, 1756

3 I must see the soldiers' bayonets levelled at
   my throat before I can be induced to give
   way.
   Robert, Lord Clive, on hearing of a
   conspiracy of British officers against his
   authority in India, 1765

4 Am I not rather deserving of praise for the
   moderation which marked my proceedings?
   Consider the situation in which the victory
   at Plassey had placed me. A great prince was
   dependent on my pleasure; an opulent city
   lay at my mercy; its richest bankers bid
   against each other for my smiles; I walked
   through vaults which were thrown open to
   me alone, piled on either hand with gold
   and jewels. Mr Chairman, at this moment I
   stand astonished at my own moderation.
   Robert, Lord Clive, statement to
   Parliament, 1773

5 Clive — that man not born for a desk —
   that heaven-born general.
   William Pitt the Elder, speech in the House
   of Commons after the Battle of Plassey,
   1757

6 A savage old Nabob, with an immense
   fortune, a tawny complexion, a bad liver,
   and a worse heart.
   Lord Macaulay, in the *Edinburgh Review*
   (1840)

7 What I like about Clive
   Is that he is no longer alive.
   There is a great deal to be said
   For being dead.
   E.C. Bentley, *Biography for Beginners*
   (1905)

35 COLLINS
   Michael, Irish revolutionary (1890–1922)

1 That volley which we have just heard is the
   only speech which it is proper to make over
   the grave of a dead Fenian.
   Michael Collins, funeral oration for
   Thomas Ashe, 1917

2 Collins had succeeded, as no Irish leader
   before him had done, in thrusting his foot
   firmly against the half-open gate of Dublin
   Castle, and England would find it
   impossible to shut it again.
   Frank O'Connor, *The Big Man* (1937)

3 Whatever else Collins was, he was a man
   who led from the front.
   M.R.D. Foot, *War and Society* (1973)

36 CORTES
   **Hernan, Spanish conquistadore
   (1485–1547)**

1 Valour loves not idleness, and so, therefore,
   if you will take hope for valour, or valour for
   hope, and if you do not abandon me, as I
   shall not abandon you, I shall make you in a
   very short time the richest of all men who
   have crossed the seas, and of all the armies
   that have here made war.
   Hernan Cortes, address to his expedition
   before beginning the conquest of Central
   Mexico, 1519, quoted in de Gomara, *Istoria
   de la Conquista de Mexico* (1552), trs.
   Simpson

2 I and my companions suffer from a disease of the heart which can be cured only with gold.
Hernan Cortes, message sent to the Lord Mcoctezuma, 1519, quoted in de Gomara, *Istoria de la Conquista de Mexico* (1552), trs. Simpson

3 There is no retreat, or, to put it more mildly, retirement, which does not bring an infinity of woes to those who make it, to wit: shame, hunger, loss of friends, goods and arms, and death, which is the worst of them, but not the last for infamy endures for ever.
Hernan Cortes, address to his army, 1520, quoted in de Gomara, *Istoria de la Conquista de Mexico* (1552), trs. Simpson

4 It will give me great pleasure to fight for my God against your gods, who are a mere nothing.
Hernan Cortes, address to the Aztec priests before the fall of Mexico, 1521, quoted in *Five Letters* (1522–5), trs. Morris

5 The Aztecs said that by no means would they give themselves up, for as long as one of them was left he would die fighting, and that we would get nothing of theirs because they would burn everything and throw it in the water.
Hernan Cortes, third despatch to the Emperor Charles V, 1522, quoted in *Five Letters* (1522–5), trs. Morris

6 He was most careful in all our campaigns, even by night, and many nights he went the rounds and challenged the sentinels, and he entered into the Ranchos and shelters of our soldiers, and if he found one without his arms and with his shoes off, he admonished him and said to him, 'To a worthless sheep, the wool seems heavy.'
Bernal Diaz del Castillo, *The Conquest of New Spain* (1568), trs. Cohen

37 CRAZY HORSE
Sioux chief (c. 1849–77)

1 We do not hunt the troops, and never have, they have always hunted us on our own ground.
Chief Crazy Horse, statement of surrender to General Nelson Miles, 1877

2 We did not ask you white men to come here.
Chief Crazy Horse, statement of surrender to General Nelson Miles, 1877

3 One does not sell the earth upon which the people walk.
Chief Crazy Horse, quoted in Brown, *Bury My Heart At Wounded Knee* (1971)

38 CROMWELL
Oliver, Lord Protector of England
(1599–1658)

1 If you chose godly, honest men to be captains of horse, honest men will follow them.
Oliver Cromwell, letter to Sir William Springe, 1643

2 Sir, the State, in choosing men to serve them, takes no notice of their opinions.
Oliver Cromwell, letter to Maj-Gen Cauford, 1644

3 On becoming soldiers we have not ceased to be citizens.
Oliver Cromwell, address to Parliament, 1647

4 (Of the Battle of Worcester) It is for aught I know a crowning mercy.
Oliver Cromwell, letter, 1651

5 Necessity hath no law.
Oliver Cromwell, speech in Parliament, 1654

6 Your poor army, these poor contemptible men, came up hither.
Oliver Cromwell, speech to the Committee of Parliament, 1657

7 Mr Lely, I desire you would use all your skill to paint my picture truly like me, and not flatter me at all; but remark all these roughnesses, pimples, warts and everything as you see me, otherwise I will never pay a farthing for it.
Oliver Cromwell, letter to Sir Peter Lely, quoted in Walpole, *Anecdotes of Painting in England* (1762)

8 *Cromwell*, our cheif of men.
John Milton, 'To the Lord Generall Cromwell, May 1652'

9 His ashes in a peaceful urn shall rest,

His name a great example stands, to show
How strangely high endeavours may be
　　blest
When piety and valour jointly go.
John Dryden, of Cromwell's death, *Heroic Stanzas* (1659)

10 He was one of those men whom his very
enemies could not condemn without
commending him at the same time.
Earl of Clarendon, *History of the Rebellion* (1704)

11 He will be looked upon by posterity as a
brave, bad man.
Earl of Clarendon, *History of the Rebellion* (1704)

12 His foreign enterprises, though full of
intrepidity were pernicious to the national
interest and seem more the result of
impetuous fury or narrow prejudices than of
cool foresight and deliberation.
David Hume, *History of England* (1754)

39　CUMBERLAND
William Augustus, Duke of Cumberland, British general (1739–76)

1 God's curse on the laws that made these
men our enemies!
Duke of Cumberland, of the Irish Brigade at
the Battle of Fontenoy, 1745, quoted in
Churchill, *A History of the English Speaking
Peoples*, vol. III (1957)

2 I am now in a country so much our enemy
that there is hardly any intelligence to be
got, and whenever we do procure any it is
the business of the country to have it
contradicted.
Duke of Cumberland, letter from Scotland
to the Duke of Newcastle, 1746

3 Lord, what am I, that I should be spared
when so many brave men lie dead upon this
spot?
Duke of Cumberland, after the Battle of
Culloden, 1746

4 And if I had my desire
the Duke would be in sad plight:
the butcher who butchered the meat
would have hemp round his throat.
Alasdair MacMhaighstir Alasdair,
'Song to the Prince'

5 (Of the Battle of Culloden, 1746)
No quarter was given on the battlefield
where Cumberland earned his long-lived
title of 'Butcher'.
Winston S. Churchill, *A History of the
English Speaking Peoples*, vol. III (1957)

40　CUSTER
George Armstrong, American general (1839–76)

1 I feel that my destiny is in the hands of the
Almighty. This belief, more than any other
facts or reason, makes me brave and fearless
as I am.
General George Armstrong Custer, letter to
his wife, 1863, quoted in Connell, *Son of the
Morning Star* (1985)

2 Hurrah, boys, we've got them!
General George Armstrong Custer, to his
troops before the Battle of Little Big Horn,
1876, quoted by Giovanni Martini in
Benteen, *The Custer Fight* (1933)

3 They say we massacred him, but he would
have massacred us had we not defended
ourselves and fought to the death.
Chief Crazy Horse, last words, 1877,
quoted in Brininstool, *Crazy Horse* (1949)

4 Where the last stand was made, the Long
Hair stood like a sheaf of corn with all the
ears fallen around him.
Chief Sitting Bull, quoted in the *New York
Herald*, 1877

5 Horse and man seemed one when the
general vaulted into the saddle.
Elisabeth Bacon Custer, *Boots and Saddles*
(1885)

41　DAYAN
Moshe, Israeli general (1915–81)

1 We cannot safeguard every water pipe and
every tree; we cannot prevent the murder of
workers in orchards or families asleep in
their beds. But we can put a high price on
our blood; too heavy a price for an Arab

settlement, an Arab army, or an Arab government to pay.
General Moshe Dayan, Order to Israeli officers, 1955

2 We are a generation of settlers, and without a helmet or gun barrel we shall not be able to plant a tree or build a house.
General Moshe Dayan, address at the funeral of Ro'i Rotberg, 1956

3 Better to be engaged in restraining the noble horse than in prodding the reluctant mule.
General Moshe Dayan, *Diary of the Sinai Campaign* (1965)

4 We must prepare ourselves morally and physically to endure a protracted struggle, not to draw up a timetable for the achievement of 'rest and peace'.
General Moshe Dayan, address to Israeli staff officers, Staff and Command School, 1969

5 Death in combat is not the end of the fight but its peak, and since combat is a part, and times the sum total of life, death which is the peak of combat, is not the destruction of life, but its fullest, most powerful expression.
General Moshe Dayan, address in honour of Natan Alterman, 1971

6 Matters of routine, discipline, training, housekeeping and general administration bored him.
Maj-Gen Chaim Herzog, *The War of Atonement* (1975)

---

42 DE GAULLE
Charles André Joseph Marie, French general (1890–1970)

1 People get the history they deserve.
General Charles de Gaulle, lecture to army cadets, St Cyr, 1920

2 International law would be worthless without soldiers to back it. Whichever way the world goes, it will not be able to do without weapons.
General Charles de Gaulle, *Le Fil de l'Epée* (1932)

3 To all Frenchmen! France has lost a battle, but France has not lost the war.
General Charles de Gaulle, proclamation, 1940

4 Whatever happens, the flame of French resistance must not, and shall not, be extinguished.
General Charles de Gaulle, radio broadcast, 1940

5 I feel not a person but an instrument of Destiny.
General Charles de Gaulle, on entering Paris, 1944

6 France cannot be France for me without grandeur. France is not France unless she is great.
General Charles de Gaulle, *Mémoires*, I (1954)

7 Men are of no importance. What counts is who commands.
General Charles de Gaulle, interview in the *New York Times*, 1968

8 No country without an atom bomb could properly consider itself independent.
General Charles de Gaulle, in the *New York Times*, 1968

9 An intelligent well-read officer, keen on his job: brilliant and resourceful qualities; a lot in him. Unfortunately he spoils his undoubted talents by his excessive assurance, his contempt for other people's point of view, and his attitude of a king in exile.
Confidential report on de Gaulle, French War College, 1922, quoted in Tournoux, *Pétain et de Gaulle* (1964)

10 The hardest cross I have to bear is the cross of Lorraine.
Winston S. Churchill, 1943, quoted in Wilson, *A Prime Minister on Prime Ministers* (1977)

11 What can you do with a man who looks like a female llama surprised when bathing?
Winston S. Churchill, in discussion with Field Marshal Viscount Alanbrooke, 1944, quoted in Fraser, *Alanbrooke* (1982)

12 One had the sense that if he moved to a window the centre of gravity might shift and the whole room might tilt everyone into the garden.
Henry Kissinger, *The White House Years* (1979)

43 EDWARD I
King of England (1239–1307)

1 To each his own.
   King Edward I, maxim adopted for himself at his coronation, 1272

2 Carry my bones before you in your march, for the rebels will not be able to endure the sight of me, dead or alive.
   King Edward I, last words addressed to his son, 1307

3 Ruin seize thee, ruthless King!
   Thomas Gray, *The Bard* (1757)

4 The model of a politic and warlike king: he possessed industry, penetration, courage, vigilance and enterprise.
   David Hume, *History of England* (1763)

5 The force and fame which Edward I had gathered in his youth and prime cast their shield over the decline of his later years.
   Winston S. Churchill, *History of the English Speaking Peoples*, vol. I (1956)

44 EDWARD III
King of England (1312–77)

1 *Honi soit qui mal y pense.* Evil be to him who evil thinks.
   King Edward III, motto adopted for the Order of the Garter, 1349

2 The Skot in his wordes has wind for to spill,
   For at the last Edward sall haue al his will.
   He had his will at Berwik, wele wurht the while!
   Skottes broght him the kayes — bot for thaire gile.
   Laurence Minot, 'On the Scots' (*c.* 1333)

3 Then folk thought that a new sun was rising over England.
   Thomas Walsingham, on King Edward III's return from France, 1347, in *Historia Anglicana*, I, ed. Riley (1863)

4 Next to God, he reposed his chief Confidence in the Valour of his own subjects.
   Joshua Barnes, *The History of King Edward III* (1688)

45 EDWARD
The Black Prince (1330–76)

1 He who is steadfast unto death shall be saved and they who suffer in a just cause, theirs is the kingdom of heaven.
   Prince Edward, address to his knights before the Battle of Poitiers, 1356, quoted in Froissart, *Chronicles* (1523–5)

2 This is the work of God, not mine; we should thank Him and pray to Him with all our heart that He may give us His grace and pardon us this victory.
   Prince Edward, letter to the Lord Mayor of London after the Battle of Poitiers, 1356, quoted in Hughes, *Illustrations of Chaucer's England* (1918)

3 Also say to them, that they suffer him this day to win his spurs, for if God be pleased, I will this journey be his, and the honour thereof.
   King Edward III of England, of the Black Prince at the Battle of Crécy, 1345, quoted in Froissart, *Chronicles* (1523–5)

4 The flower of the world's knighthood at that time and the most successful soldier of his age.
   Jean Froissart, *Chronicles* (1523–5)

5 In war, was never lion rag'd more fierce,
   In peace was never gentle lamb more mild,
   Than was that young and princely gentleman.
   William Shakespeare, *Richard II*, II, i

46 EISENHOWER
Dwight David, American general and President of the United States (1890–1969)

1 The question is just how long can you hang this operation on the end of a limb and let it hang there?
   General Dwight D. Eisenhower, giving the Order for the commencement of the D-day operations, 1944

2 It wearies me to be thought of as timid,

when I've had to do things that were so risky as to be almost crazy.
General Dwight D. Eisenhower, Diary, 1944

3 The eyes of the world are upon you. The hopes and prayers of liberty-loving people everywhere march with you.
General Dwight D. Eisenhower, Order to his troops on D-Day, 1944

4 In the organisation and composition of my staff, we proceeded as if all its members belonged to a single nation.
General Dwight D. Eisenhower, *Crusade in Europe* (1948)

5 Nothing is easy in war. Mistakes are always paid for in casualties and troops are quick to sense any blunders made by their comrades.
General Dwight D. Eisenhower, *Crusade in Europe* (1948)

6 In the final choice a soldier's pack is not so heavy a burden as a prisoner's chains.
General Dwight D. Eisenhower, Inaugural Address as President, 1953

7 I feel compelled to speak today in a language that in some sense is new — one which I, who have spent so much of my life in the military profession, would have preferred not to use. That new language is the language of atomic warfare.
General Dwight D. Eisenhower, address to the United Nations General Assembly, 1953

8 Peace and justice are two sides of the same coin.
General Dwight D. Eisenhower, news conference at the United Nations, New York, 1957

9 There can be no such thing as Fortress America. If ever we reduced to the isolation implied by that term we would occupy a prison, not a fortress.
General Dwight D. Eisenhower, State of the Union Address, 1959

10 Universally trusted, he evoked spontaneous affection, respect and loyalty from political and military leaders alike.
Chester Wilmot, *The Struggle for Europe* (1952)

11 A remarkable and most loveable man.
Field Marshal Viscount Montgomery, *Memoirs* (1958)

12 His was an excellent appointment, and he

carried out his assignment with great distinction.
Field Marshal Earl Alexander, *Memoirs* (1962)

EL CID
See 125. RODRIGO DIAZ DE VIVAR

---

47 EUGENE
Prince of Savoy-Carignon, Imperial soldier (1663–1736)

1 I wish to fight among brave men and not among cowards.
Prince Eugene of Savoy, Order rallying the Imperial Cavalry at the Battle of Blenheim, 1704

2 Money, which you don't want in England, will buy fine clothes and fine horses, but it can't buy that lively air I see in every one of these troopers' faces.
Prince Eugene of Savoy, on inspecting the Duke of Marlborough's cavalry, 1704

3 He who has not seen this has seen nothing.
Prince Eugene of Savoy, after the Battle of Oudenarde, 1708

4 I am afraid that we must expect things to go from bad to worse in England so long as a woman is in charge.
Prince Eugene of Savoy, letter to the Imperial ambassador in The Hague, 1710

5 I know how to hold my tongue when it is necessary; but I do not know how to lie.
Prince Eugene of Savoy, to the English diplomat Lord Waldegrave, 1728

6 The common soldiers should not be exhausted except with good reason, and you should only be harsh when, as often happens, kindness proves useless.
Prince Eugene of Savoy, letter to General Count Otto von Traun, 1728

7 It is part of Eugene's character to see difficulties and obstacles before beginning a task. When the moment for action arrives, then he is all strength and activity.
Count John von Schulenburg, 1704, quoted in Frischauer, *Prince Eugene* (1934)

8   He lets no pleasures disturb his noble
    ambition and he is a Mars without Venus, at
    least at the present time.
    E.G. Rinck, *Life of Leopold I* (1708)

48 FAIRFAX
Thomas, Baron Fairfax, British soldier
(1612–71)

1   The Crown of England is and will be where
    it ought to be; we fight to maintain it there.
    Thomas, Lord Fairfax, letter to Prince
    Rupert during the Siege of Bristol, 1643,
    quoted in Sprigge, *Anglia Rediviva* (1647)

2   Human probabilities are not sufficient
    grounds to make war upon a neighbour
    nation.
    Thomas, Lord Fairfax, letter to Thomas
    Harrison refusing to take part in the
    invasion of Scotland, 1650, quoted in
    Slingsby, *Diary* (1836)

3   A setter up and putter down of parliaments.
    Denzil Holles, on Fairfax's entry into
    London, 1647, quoted in Maseres, *Select
    Tracts Relating to the Civil War in England*
    (1815)

4   *Fairfax*, whose name in armes through
                Europe rings.
    John Milton, 'On the Lord Gen. Fairfax at
    the seige of Colchester' (1648)

5   It hath pleased the Lord of Hosts, who was
    called upon to decide the controversy of this
    nation, to write His name upon your sword
    in very legible characters.
    Colonel Thomas Pride, Petition to Fairfax,
    1648, quoted in *The Moderate Intelligencer*
    (1648)

49 FALKENHAYN
Erich von, German general (1861–1922)

1   The first principle in position warfare must
    be to yield not one foot of ground; and if it
    be lost to re-take it by immediate
    counter-attack, even to the use of the last
    man.
    General Erich von Falkenhayn, Order to the
    German Second Army before the Battle of

the Somme, 1916, quoted in Edmonds,
Military Operations, France and Belgium,
vol. II (1922–32)

2   Rhetoric, self-adulation and lies plunged
    Germany into the deepest abyss when they
    stifled the sense of reality in our once strong
    and good people.
    General Erich von Falkenhayn, preface to
    General Headquarters 1914–1916 (1919)

3   The soldier who is well-disciplined and has
    his heart in the business, and in addition has
    learned to attack, is equal to any situation in
    war.
    General Erich von Falkenhayn, *General
    Headquarters 1914–1916* (1919)

4   Falkenhayn is the evil genius of our
    Fatherland, and, unfortunately, he has the
    Kaiser in his pocket.
    General Max von Hoffmann, on the
    appointment of General von Falkenhayn as
    Chief of General Staff, Diary, 1914

5   The only general on either side in the First
    World War who aspired to something less
    than decisive victory.
    A.J.P. Taylor, *The First World War* (1963)

50 FOCH
Ferdinand, Marshal of France
(1851–1929)

1   I have only one merit: I have forgotten what
    I taught and what I learned.
    Marshal Ferdinand Foch, 1914, quoted in
    Monteilhet, *Les Institutions Militaires de la
    France* (1932)

2   (Of the Battle of the Marne, 1914) My
    centre is giving way, my right is in retreat;
    situation excellent. I shall attack.
    Marshal Ferdinand Foch, quoted in Aston,
    Biography of Foch (1929)

3   *L'édifice commence à craquer. Tout le
    monde à la bataille.* The edifice begins to
    crumble. Everyone advance.
    Marshal Ferdinand Foch, despatch to
    British GHQ, 1918

4   Your greatness does not depend upon the
    size of your command, but on the manner in
    which you exercise it.
    Marshal Ferdinand Foch, quoted in Aston,
    Biography of Foch (1929)

5   Action is the governing rule of war.
    Marshal Ferdinand Foch, *Precepts* (1919)

6   The most solid moral qualities melt away
    under the effect of modern arms.
    Marshal Ferdinand Foch, *Precepts* (1919)

7   It is in his eyes and the expression of his face
    that one sees his extraordinary power.
    Field Marshal Sir John French, *1914* (1919)

8   Death, which knows no nationality, allows
    an old enemy to lower his sword before
    Marshal Foch, who was a great soldier and
    a great Frenchman.
    General Hans von Seeckt, quoted in Aston,
    Biography of Foch (1929)

---

51 FRANCIS I
    King of France (1494–1547)

---

1   Let him who loves me, follow me.
    King Francis I, Order to his officers who
    opposed him fighting the Battle of
    Marignano, 1515

2   For my honour, and the honour of my
    country, I prefer honest imprisonment to
    shameful flight.
    King Francis I, letter to the Royal Escort
    and the Knights of the Realm, after losing
    the Battle of Pavia, 1525

3   Out of all I had, only honour remains, and
    my life, which is safe.
    King Francis I, letter to his mother after
    losing the Battle of Pavia, 1525

---

52 FRANCO
    **Francisco, Spanish general and head of
    state (1892–1975)**

---

1   Death in battle is the highest honour. One
    dies only once. Death comes without pain.
    To die is not as terrible as it seems. The
    most horrible thing is to live a coward.
    General Francisco Franco, 'The Legion's
    Creed', 1920, quoted in Bolin, *Spain: The
    Vital Years* (1967)

2   In war the heart must be sacrificed.
    General Francisco Franco, *Diario de una
    Bandera* (1922)

3   Men, not materials, are the dearest
    commodity in war.
    General Francisco Franco, *Diario de una
    Bandera* (1922)

4   The Army, calm and united, must sacrifice
    all personal thought and all ideology for the
    good of the nation and the tranquillity of the
    motherland.
    General Francisco Franco, General Order to
    the Spanish Army, 1931

5   War was my job; I was sure of that.
    General Francisco Franco, 1938, quoted in
    Galinsoga, *Centinela de Occidente* (1956)

6   Lord, graciously accept the effort of this
    people which always was Thine and which
    with me and in Thy Name has with heroism
    defeated the enemy of truth of this age.
    General Francisco Franco, Prayer of
    Thankgsgiving on the conclusion of the
    Spanish Civil War, Madrid, 1939

7   The beautiful page of history that you are
    writing with your lives and blood is a living
    example of what men can do who stake
    everything in the fulfilment of duty.
    King Alfonso XIII of Spain, letter to
    General Franco following the Moroccan
    campaign, 1925

8   I say there is far more liberty in Spain under
    General Franco than in any of the countries
    behind the Iron Curtain.
    Winston S. Churchill, speech in House of
    Commons, 1948

9   Caesar and Pompey, Brutus and Antony,
    Cato and Cicero — all, with all their genius,
    lacked the minor talent of being able to
    survive: Franco was the Octavius of Spain.
    Hugh Thomas, *The Spanish Civil War*
    (1961)

---

53 FREDERICK I
    **('Frederick Barbarossa'), Holy Roman
    Emperor (1122–90)**

---

1   Let it be known to the entire world that
    although we are clothed in the dignity and
    glory of the Roman Empire, this dignity
    does not keep us from human error;
    imperial majesty does not preserve us from
    ignorance.
    Frederick I, Barbarossa, Oath on signing
    the Peace of Venice, 1177

2 An Emperor is subject to no one but God
and Justice.
Frederick I, Barbarossa, quoted in Zincgref,
Apophthegmata, I (1626)

3 It pleases him to have his camp display the
panoply of Mars rather than of Venus.
Bishop Otto of Freising, *Gesta Friderici
Imperatoris* (*c.* 1150)

9 The mob, which everywhere is the majority,
will always let itself be led by scoundrels.
King Frederick II, letter to Jean Rond
d'Alembert, 1782

10 He fiddles and fights as well as any man in
Christendom.
Voltaire, letter to Sir Everard Fawkenden,
1742

54 FREDERICK II
('Frederick the Great'), King of Prussia
(1713–86)

1 I approve of all methods of attacking
provided they are directed at the point
where the enemy's army is weakest and
where the terrain favours them least.
King Frederick II, *Instructions for his
Generals* (1747)

2 If you wish to be loved by your soldiers, do
not lead them to slaughter.
King Frederick II, *Instructions for his
Generals* (1747)

3 In the face of the storm and the threat of
shipwreck, I must think, live and die like a
king.
King Frederick II, letter to Voltaire, 1757

4 Only the worthy get killed; my type always
survives.
King Frederick II, letter to his sister
Wilhelmina, 1757

5 Should I be taken prisoner by the enemy, I
forbid the smallest consideration for my
person or the least notice to be taken of
anything I write from captivity.
King Frederick II, letter to Count
Finckenstein, 1757

6 There is little glory in defeating me.
King Frederick II, letter to Lord George
Keith, 1757

7 You rogues, do you want to live for ever?
King Frederick II, to the Prussian Guards
when they hesitated at the Battle of Kolin
(1757)

8 Don't forget your great guns, which are the
most respectable arguments of the right of
kings.
King Frederick II, letter to his brother, 1759

55 FRENCH
**John Denton Pinkstone, Earl of Ypres,
British field marshal (1852–1925)**

1 Have captured forty engines, seventy
wagons of stores, eighty women all in good
working order.
Field Marshal Sir John French, despatch to
Field Marshal Lord Roberts, 1900

2 I think the battle of the Aisne is very typical
of what battles in the future are most likely
to resemble. Siege operations will enter
largely into the tactical problems — the
spade will be as great a necessity as the rifle,
and the heaviest calibres and types of
artillery will be brought up in support of
either side.
Field Marshal Sir John French, letter to
King George V, 1914

3 How I should love to have a real good 'go'
at them in the open with lots of cavalry and
horse artillery and run them to earth.
Field Marshal Sir John French, letter to
Winifred Bennett, 1915

4 War is really a very brutal way of settling
differences and the more I see of it the more
I hate it.
Field Marshal Sir John French, letter to
Winifred Bennett, 1915

5 It is a solemn thought that at my signal all
these fine young fellows go to their death.
Field Marshal Sir John French, quoted in
Brett, *The Journals and Letters of Reginald,
Viscount Esher* (1934)

6 My personal experiences have showed me
the great necessity for a General in
command during an extensive engagement
to keep in the most commanding position
and in one place, if possible.
Field Marshal Sir John French, quoted in
French, *War Diaries* (1937)

7   To establish a 'moral superiority' over the
    enemy is an object of the *first importance*.
    Field Marshal Sir John French, quoted in
    French, *War Diaries* (1937)

8   French is the most thoroughly loyal,
    energetic soldier I have, and all under him
    are devoted to him — not because he is
    lenient, but because they admire his
    soldier-like qualities.
    Field Marshal Earl Kitchener, letter to Field
    Marshal Lord Roberts, 1902

9   Sir John French may not have been a great
    soldier in the modern sense of the word, but
    he was a great leader of men.
    C.D. Baker-Carr, *From Chauffeur to
    Brigadier-General* (1930)

10 To no one of her many distinguished sons
    does our Empire owe a greater debt of
    gratitude.
    Field Marshal Viscount Allenby, quoted in
    French, *Life of Field Marshal Sir John French*
    (1931)

11 French was a natural soldier.
    Winston S. Churchill, *Great
    Contemporaries* (1937)

12 Johnnie French aroused anything but
    indifference amongst his contemporaries.
    Richard Holmes, *The Little Field-Marshal*
    (1981)

56 GARIBALDI
Giuseppe, Italian revolutionary (1807–82)

1   I offer neither pay, nor quarters, nor food; I
    offer only hunger, thirst, forced marches,
    battle and death. Let him who loves his
    country with his heart, and not merely with
    his lips, follow me.
    Giuseppe Garibaldi, address to his besieged
    troops in Rome, 1849

2   One hour of our life in Rome is worth a
    century of ordinary existence.
    Giuseppe Garibaldi, letter to his wife after
    the fall of Rome, 1849

3   Soldiers, I release you from your duty to
    follow me, and leave you free to return to
    your homes. But remember that although
    the Roman war for the independence of

Italy has ended, Italy remains in shameful
slavery.
Giuseppe Garibaldi, Order of the Day at the
conclusion of the Roman campaign, 1849

4   Here we shall make Italy — or die!
    Giuseppe Garibaldi, Order to his troops
    before the Battle of Calatafimi, 1860

5   I absolutely forbid anyone to get wounded.
    Giuseppe Garibaldi, Order to his troops
    before the taking of Reggio Calabria, 1860

6   You have a right to exult this day. It is the
    beginning of a new epoch not only for you
    but for the whole of Italy, of which Naples
    forms the fairest part.
    Giuseppe Garibaldi, address to his troops
    after the fall of Naples, 1860

7   The tombs are uncovered, the dead come
                    from afar,
    The ghosts of our martyrs are rising to war,
    With swords in their hands, and with laurels
                    of fame,
    And dead hearts still glowing with Italy's
                    name.
    Anon., 'Garibaldi's Hymn', sung by the
    Garibaldini, 1859

57 GENGHIS KHAN
Mongol leader (c. 1162–1227)

    The merit of an action lies in finishing it to
    the end.
    Genghis Khan, advice to his sons on being
    acclaimed Khan of Khans, 1206

2   God in Heaven, and Genghis on Earth,
    Khan by the power of God and Emperor of
    all men.
    Genghis Khan, motto adopted for his
    imperial seal, 1221

3   The greatest pleasure is to vanquish your
    enemies, to chase them before you, to rob
    them of their wealth, to see their near and
    dear bathed in tears, to ride their horses
    and sleep on the white bellies of their wives
    and daughters.
    Genghis Khan, advice to his generals, after
    invading Russia, 1224, quoted in Matthew
    Paris, *Historia Majora Matthei Paris
    Monachi Albanensis* (1571)

4 My life was too short to achieve the
conquest of the whole world.
Genghis Khan, last words to his sons,
1227, quoted in *The Secret History of the
Mongols* (1242)

5 He had acquired the invariable habit of
conscripting the soldiers of a conquered
army into his own, with the object of
subduing other countries by virtue of his
increasing strength, as is clearly evident in
his successors, who imitate his wicked
cunning.
Benedict the Pole, *Tartar Relation,*
(*c.* 1240), quoted in Skelton, Marston and
Painter, *The Vinland Map and the Tartar
Relation* (1965)

58 GEORGE II
King of Great Britain (1683–1760)

1 Now boys! Now for the honour of England!
Fire, and behave brave and the French will
run!
King George II, Order to his troops before
the Battle of Dettingen, 1743

2 Nor so did behave
Young Hanover brave
In this bloody field, I assure ye.
When his war-horse was shot,
He valued it not,
But fought it on foot like a fury.
Jonathan Swift, on George's behaviour at
the Battle of Oudenarde, 1708, in 'Jack
Frenchman's Lamentation'

3 His character would not afford subject for
epic poetry, but will look well in the sober
page of history.
Elizabeth Montague, on the death of King
George II, 1760

4 He has as much bravery as any man, though
his political courage seems somewhat
problematical: however, it is a fault on the
right side; for had he always been as firm
and undaunted in the closet as he shewed
himself at Oudenarde and Dettingen, he
might not have proved quite so good a king
in this limited monarchy.
Lord Waldegrave, *Memoirs 1754–1758*
(1821)

5 It was said of King George II that he fought

with the sword in one hand and the lash in
the other.
Marjorie Ward, *The Blessed Trade* (1971)

59 GERONIMO
Apache chief (d. 1909)

1 Once I moved about like the wind. Now I
surrender to you and that is all.
Chief Geronimo, statement of surrender to
General George Crook, 1886

2 How long will it be until it is said there are
no Apaches?
Chief Geronimo, quoted in Barrett,
Geronimo: His Own Story (1906)

3 The Indians always tried to live peaceably
with the white soldiers and settlers.
Chief Geronimo, quoted in Barrett,
Geronimo: His Own Story (1906)

4 I don't fight Mexicans with cartridges. I
fight them with rocks and keep my
cartridges to fight the white soldiers.
Chief Geronimo, quoted in Davis, *The
Truth about Geronimo* (1929)

60 GIAP
**Vo Nguyen, Vietnamese general
(1910–   )**

1 We will have to receive aid from abroad in
order to be able to carry out the
counter-offensive, but to count solely upon
it without taking into account our own
capabilities is to show proof of subjectivism
and of lack of political conscience.
General Vo Nguyen Giap, lecture to
political commissars, Indochina, 1950

2 We made too many deviations and executed
too many honest people. We attacked on
too large a front and, seeing enemies
everywhere, resorted to terror, which
became far too widespread.
General Vo Nguyen Giap, address to the
Tenth Congress of the Vietnam Communist
Party Central Committee, 1960

3 If the Vietnamese people's war of liberation
ended in a glorious victory, it is because we

did not fight alone, but with the support of progressive peoples the world over, and more especially the peoples of the brother countries with the Soviet Union at the head.
General Vo Nguyen Giap, *People's War — People's Army* (1961)

4   The fundamental principle of revolutionary war: strike to win, strike only when success is certain; if not, then don't strike.
General Vo Nguyen Giap, *People's War — People's Army* (1961)

5   The People's army is the instrument of the Party and of the revolutionary State for the accomplishment, in the armed form, of the tasks of the revolution.
General Vo Nguyen Giap, *People's War — People's Army* (1961)

6   Revolutionary armed struggle in any country has common fundamental laws. Revolutionary armed struggle in each country has characteristics and laws of its own too.
General Vo Nguyen Giap, *People's War — People's Army* (1961)

GLENDOWER
See 108. OWAIN GLYN DWR

---

61 GNEISENAU
August Wilhelm, Graf Neithardt von Gneisenau, Prussian field marshal (1760–1831)

---

1   In times of peace we have neglected much, occupied ourselves with frivolities, flattered the people's love of shows and neglected war.
Field Marshal August von Gneisenau, on the mobilization of Prussia, 1806, quoted in Seeley, *Life and Times of Stein* (1878)

2   To the man of honour nothing is left but to envy those who fall in the field of battle.
Field Marshal August von Gneisenau, after the Battle of Auerstadt, 1806, quoted in Seeley, *Life and Times of Stein* (1878)

3   The shame of an army annihilated through

misfortunes due to its own fault can never be wiped away.
Field Marshal August von Gneisenau, after the capitulation of the Prussian army to Marshal Bernadotte, 1806, quoted in Henderson, *Blücher and the Uprising against Napoleon* (1911)

4   Religion, prayer, love of one's ruler, love of the Fatherland — these things are nothing less than poetry.
Field Marshal August von Gneisenau, memorandum to King Frederick William III, 1811

5   We must take revenge for the many sorrows inflicted upon the nation, and for so much arrogance. If we do not, then we are miserable wretches indeed, and deserve to be shocked out of our lazy peace every two years and threatened with the scourge of slavery.
Field Marshal August von Gneisenau, letter to Heinrich von Stein, 1814, quoted in Parkinson, *Clausewitz* (1970)

6   Gneisenau makes the pills which I administer.
Marshal Blücher, after the Battle of Leipzig, 1813

---

62 GORDON
Charles George, British general (1833–85)

---

1   I would sooner live like a Dervish with the Mahdi, than go out to dinner every night in London.
General Charles George Gordon, Khartoum journal, 1883

2   I know if *I* was chief, I would never employ myself, for I am incorrigible.
General Charles George Gordon, Khartoum journal, 1883

3   It is not the fear of death, that is past, thank God, but I fear defeat and its consequences. I do not believe in the calm, unmoved man.
General Charles George Gordon, Khartoum journal, 1884

4   When God was portioning out fear to all the people in the world, at last, it came to my turn, and there was no fear left to give me. Go tell all the people in Khartoum that Gordon fears nothing, for God has created him without fear.

General Charles George Gordon,
address to the people of Khartoum,
Christmas Day, 1884, quoted in the diary of
Bordeini Bey, 1884

5 NOW MARK THIS, if the Expeditionary
Force, and I ask for no more than 200 men,
does not come in ten days, *the town may
fall*; and I have done my best for the honour
of my country.
General Charles George Gordon, last entry
in his Khartoum journal, 1884

6 Khartoum is all right. Could hold out for
years.
General Charles George Gordon, last
message to General Wolseley's Relief Force,
1884

7 Too late! Too late to save him.
In vain, in vain they tried.
His life was England's glory.
His death was England's pride.
Anon., popular song commemorating the
death of General Charles Gordon, 1885

8 Never was a garrison so nearly rescued,
never was a Commander so sincerely
lamented.
Field Marshal Earl Kitchener, 'Notes on the
Fall of Khartoum', 1885

9 (Of Gordon's death) Grief inexpressible!
Queen Victoria, letter to Augusta Gordon,
1885

10 A man who habitually consults the prophet
Isaiah when he is in difficulty is not apt to
obey the orders of anyone.
Sir Evelyn Baring, quoted in Trench,
Charley Gordon (1978)

11 His reading was confined almost entirely to
the Bible; but the Bible he read and re-read
with an untiring and an unending assiduity.
There, he was convinced, all truth was to be
found, and he was equally convinced that he
could find it.
Lytton Strachey, *Eminent Victorians* (1918)

63 GRAHAM

**John Graham of Claverhouse, Viscount
Dundee, Jacobite general (1648–89)**

1 I am sure whatever evill befall the country,
the King is innocent, and I have done my
deuty. I need tell yow no news; yow know
all better then I doe, who dwell in deserts.

John Graham of Claverhouse, letter to the
Duke of Atholl, enlisting his support for
King James II and VII, 1689

2 If you have a mynd to preserve yourself and
save the King, be in armes to morou, that
when the letter comes you may be here in a
day. All the world will be with us, blissed be
God.
John Graham of Claverhouse, letter to
Duncan Macpherson of Cluny before the
Battle of Killiecrankie, 1689

3 'Tis the less matter for me, seeing the day
goes well for my master.
John Graham of Claverhouse, dying words
after his victory at the Battle of
Killiecrankie, 1689

4 To the Lords of Convention, 'twas Claver'se
who spoke,
''Ere the King's crown shall fall there are
crowns to be broke;
So let each Cavalier who loves honour and
me
Come follow the bonnet of Bonny Dundee.'
Sir Walter Scott, 'Bonny Dundee', *The
Doom of Devorgoil* (1830)

5 Last of Scots, and last of freemen —
Last of all that dauntless race,
Who would rather die unsullied
Than outlive the land's disgrace!
W.E. Aytoun, 'The Burial March of
Dundee', *Lays of the Scottish Cavaliers*
(1849)

64 GRANT

**Ulysses Simpson, American (Union)
general and President of the United States
(1822–85)**

1 No terms except an unconditional and
immediate surrender can be accepted.
General Ulysses S. Grant, ultimatum to
General Buckner at Fort Donelson, 1862

2 I propose to fight it out on this line if it takes
all summer.
General Ulysses S. Grant, dispatch to
General Henry W. Walleck, 1864

3 Your men must keep their horses and mules.
They will need them for the spring
ploughing.
General Ulysses S. Grant, Order to General
Lee at the end of the American Civil War,
1865

4 Laws are to govern all alike — those

opposed as well as those who favour them. I know of no method to repeal of bad or obnoxious laws so effective as their stringent execution.
General Ulysses S. Grant, Inaugural Address as President, 1869

5 My object in war was to exhaust Lee's army. I was obliged to sacrifice men to do it. I have been called a butcher. Well, I never spared lives to gain an object; but then I gained it, and I knew it was the only way.
General Ulysses S. Grant, quoted in Chancellor, *An Englishman in the American Civil War* (1971)

6 Grant still represents the butcher type of general.
Maj-Gen J.F.C. Fuller, *The Generalship of Ulysses S. Grant* (1929)

7 The modernity of Grant's mind was most apparent in his grasp of the concept that war was becoming total and that the destruction of the enemy's resources was as effective and legitimate a form of warfare as the destruction of his armies.
T. Harry Williams, *Lincoln and His Generals* (1952)

65 GUDERIAN
Heinz, German general (1888–1953)

1 Until our critics can produce some new and better method of making a successful land attack other than self-massacre, we shall continue to maintain our belief that tanks — properly employed, needless to say — are today the best means available for a land attack.
Colonel-General Heinz Guderian, *Achtung! Panzer!* (1937)

2 Tanks are a life-saving weapon.
Colonel-General Heinz Guderian, remark to Adolf Hitler, 1939, quoted in *Panzer Leader* (1952)

3 The icy cold, the lack of shelter, the shortage of clothing, the heavy losses of men and equipment, the wretched state of our fuel supplies, all this makes the duties of a commander a misery, and the longer it goes

on the more I am crushed by the enormous responsibility I have to bear.
Colonel-General Heinz Guderian, letter to his wife during the German advance to Moscow, 1941

4 We have severely underestimated the Russians, the extent of the country and the treachery of the climate. This is the revenge of reality.
Colonel-General Heinz Guderian, letter to his wife during Operation Barbarossa, 1941

5 The Eastern Front is like a house of cards. If the front is broken through at one point all the rest will collapse.
Colonel-General Heinz Guderian, interview with Adolf Hitler, 1944

6 Your operations always hang by a thread.
Field Marshal Günther von Kluge, before the crossing of the River Dnieper by Guderian's 2nd Panzer Army, 1941

66 GUEVARA
Ernesto 'Che', Cuban guerrilla leader (1928–67)

1 One does not necessarily have to wait for a revolutionary situation: it can be created.
Ernesto 'Che' Guevara, *Guerrilla Warfare* (1961)

2 Discipline and morale are the foundations on which the strength of an army rests, whatever its composition.
Ernesto 'Che' Guevara, *Reminiscences of the Cuban Revolutionary War* (1968)

3 The peasant fought because he wanted land for himself, for his children, to manage it, sell it, and get rich by his work.
Ernesto 'Che' Guevara, *Reminiscences of the Cuban Revolutionary War* (1968)

4 Revolution cleanses men, improving them as the experimental farmer corrects the defects of his plant.
Ernesto 'Che' Guevara, *Reminiscences of the Cuban Revolutionary War* (1968)

5 Military discipline makes me vomit.
Ernesto 'Che' Guevara, quoted in *La Republica,* 1968

6 We must proceed along the path of

liberation, even if that costs millions of atomic victims.
Ernesto 'Che' Guevara, quoted in *Verde Olivo*, 1968

7 Guevara was a powerful theoretician but no soldier.
Shelford Bidwell, *Modern Warfare* (1973)

67 GUSTAVUS ADOLPHUS
King of Sweden (1594–1632)

1 If I draw a pail of water from the Baltic, am I supposed to be desirous of drinking up the whole sea?
King Gustavus Adolphus, before the invasion of Livonia, 1626, quoted in Ahnlund, *Axel Oxenstierna intill Gustav Adolfs död* (1940)

2 If we cannot say: *Bellum se ipsum alet* [war must pay for itself], I see no way out of all that we have engaged in.
King Gustavus Adolphus, letter to Count Oxenstierna after invading Prussia, 1628, quoted in Ahnlund, *Axel Oxenstierna intill Gustav Adolfs död* (1940)

3 A good Christian will never make a bad soldier.
King Gustavus Adolphus, after landing at Usedom, 1630 quoted in *The Swedish Intelligencer* (1632)

4 Now in very truth I believe that God has delivered him into my hands.
King Gustavus Adolphus, before engaging Wallenstein's army at the Battle of Lützen, 1632

5 What is the use of a king in a box?
King Gustavus Adolphus, before his death at the Battle of Lützen, 1632

6 The Captain of Kings and the King of Captains.
Captain Sir Robert Munro, *Recollections of a Scots Officer of a Worthy Scots Regiment with Gustavus Adolphus* (1637)

7 His mind was such that he could never believe anything was done well unless he did it himself.
Captain Sir Robert Munro, *Recollections of a Scots Officer of a Worthy Scots Regiment with Gustavus Adolphus* (1637)

8 Consider the great Gustavus Adolphus! In eighteen months he won one battle, lost a second, and was killed in the third. His fame was won at a bargain price.
Napoleon I, letter to General Gaspard Gourgaud, 1818

68 HAIG
Douglas, Earl Haig of Bemersyde, British field marshal (1861–1928)

1 It behoves everyone to do his little and try and qualify for as high a position as possible. It is not ambition. This is *duty*.
Field Marshal Earl Haig, letter to his nephew, 1902

2 It is a good thing to see the inside of the War Office for a short time, as it prevents one from having any respect for an official letter, but it is a mistake to remain there too long.
Field Marshal Earl Haig, on leaving the War Office, 1909, quoted in Marshall-Cornwall, *Haig as Military Commander* (1973)

3 I congratulate you on your running. You have run well, I hope you will run as well in the presence of the enemy.
Field Marshal Earl Haig, on presenting prizes at Aldershot garrison, 1912, quoted in Charteris, *Field Marshal Earl Haig* (1929)

4 I can honestly say that no Commander in our Army has been so highly tested as I have been in this war.
Field Marshal Earl Haig, letter to his wife, 1915

5 The enemy should never be given a complete rest by day or by night, but be gradually and relentlessly worn down by exhaustion and loss until his defence collapses.
Field Marshal Earl Haig, quoted in Edmonds, *Military Operations 1915* (1922)

6 With more guns and ammunition and more troops, the Allies were bound in the end to defeat the Germans and break through.
Field Marshal Earl Haig, Diary, 1915

7 In my opinion, it is much less costly in lives to keep on pressing the enemy after several victorious battles than to give him time to recover and organise a fresh line of defence.
Field Marshal Earl Haig, Diary, 1918

8 Every position must be held to the last man:
there must be no retirement. With our backs
to the wall, and believing in the justice of
our cause, each one of us must fight on to
the end.
Field Marshal Earl Haig, Order of the Day
to the British troops, 12 April 1918

9 He came to regard himself with almost
Calvinistic faith as the predestined
instrument of Providence for the
achievement of victory for the British
Armies.
Brig-Gen John Charteris, *Field Marshal
Earl Haig* (1929)

10 Haig was a dour Scotsman and the dullest
dog I ever had the happiness to meet.
Lord Chetwode, letter to Lord Wigram,
1935

11 Haig was a painstaking soldier with a sound
intelligence of secondary quality.
David Lloyd George, *War Memoirs* (1936)

12 If there are some who would question
Haig's right to rank with Wellington in
British military annals, there are none who
will deny that his character and conduct as a
soldier and subject will long serve as an
example to all.
Winston S. Churchill, *Great
Contemporaries* (1937)

13 In his qualities and defects he was the very
embodiment of the national character and
the army tradition.
Captain Sir Basil Liddell Hart, *Through the
Fog of War* (1938)

14 His was the only army of the great nations
at war which did not break.
Maj-Gen sir John Kennedy, in *The
Scotsman,* 1959

15 He had such a vast number of troops under
his command and was so completely remote
from the actual fighting that he was merely a
name, a figurehead.
George Coppard, *With a Machine Gun to
Cambrai* (1969)

---

69 HANNIBAL
Carthaginian general (247–183 BC)

---

1 A despised enemy has often fought a bloody

battle and famous states have often been
captured by a very slight effort.
Hannibal, address to his army before the
Battle of Ticino, 218 BC, quoted in Livy,
Histories

2 For you the prize of victory is not to possess
horses and cloaks, but to be the most envied
of mankind, masters of all the wealth of
Rome.
Hannibal, address to his army after the
Battle of Ticino, 218 BC, quoted in Livy,
Histories

3 I am not carrying on a war of extermination
against the Romans. I am contending for
honour and empire.
Hannibal, address to Roman prisoners after
the Battle of Cannae, 216 BC, quoted in
Livy, *Histories*

4 Let us now put an end to the life which has
caused the Romans so much anxiety.
Hannibal, last words after taking poison,
183 BC

5 In very truth the gods bestow not on the
same all their gifts; you know how to gain a
victory, Hannibal; you know not how to use
one.
Maharbal, on Hannibal's failure to take
Rome after the Battle of Cannae, 216 BC

6 To his reckless courage in encountering
dangers he united the greatest judgement
when in the midst of them.
Livy, *Histories*

7 Seeing the result of his work thus fast
ripening, Hannibal sat quietly on the
summit of Tifata, to break forth like the
lightning flash when the storm should be
fully gathered.
Thomas Arnold, *The Second Punic War*
(1838)

---

70 HENRY IV
King of France (1553–1610)

---

1 I am your King. You are Frenchmen. There
is the enemy. Charge!
King Henry IV, Order to his army at the
Battle of Ivry, 1590

2 Hang yourself, brave Crillon; we fought at
Arques and you were not there.
King Henry IV, letter of 1597

3 Let my white panache be your rallying point.
King Henry IV, attr. battle cry

4 Navarre shall be the wonder of the world.
William Shakespeare, *Love's Labour Lost*, I, i

5 Press where ye see my white plume shine,
Amidst the ranks of war,
And be your oriflamme today the
Helmet of Navarre.
Lord Macaulay, 'Ivy'

---

71 HENRY V
King of England (1387–1422)

---

1 We exhort you in the bowels of Jesus Christ to execute and do that thing that the Evangelist teacheth, saying, 'Friend, pay that thou owest and restore that thou wrongfully detainest.' And to the end that the blood of innocence be not spilt, we require due restitution of our rightful inheritance by you wrongfully withholden from Us.
King Henry V, letter to the Dauphin of France demanding the return of Flanders and Aquitaine to England, 1415

2 Even if our enemies enlist the greatest armies, my trust is in God, and they shall not hurt my army nor myself. I will not allow them, puffed up with pride, to rejoice in misdeeds, nor, unjustly against God, to possess my goods.
King Henry V, address to his Council before Harfleur, 1415

3 Sirs and fellows, as I am true knight and king, for me this day shall never England ransom pay.
King Henry V, address to his army before the Battle of Agincourt, 1415

4 Thou liest, thou liest, my portion lies with the Lord Jesus Christ.
King Henry V, words on his deathbed, 1422

5 He was a prince of distinguished appearance and commanding stature; and although his expression seemed to hint at pride he nevertheless made it a point of honour to

treat everybody of no matter what rank or degree with the utmost affability.
Chronique de Réligieux de Saint-Denys (1380–1422)

6 It was by his hand, as though beneath the scourge of God, that the noble blood of France was so piteously shed at Agincourt.
Georges Chastellain, chronicle of 1415, quoted in de Lettenhove, *Oeuvres* (1863–6)

7 No tongue can tell of his renown, yet if you rank all princes according to their worth, then place King Henry at their head.
John Page, *The Siege of Rouen* (1421), trs. Hutchison

8 Passing the bounds of modesty, he was a fervent soldier of Venus as well as of Mars; youthlike he was fired by her torches.
Gesta Henrici Quinti (The Deeds of Henry V) (*c.* 1425)

9   Every wretch, pining and pale before,
Beholding him, plucks comfort from his looks . . .
A little touch of Harry in the night.
William Shakespeare, *Henry V*, IV, Prologue

10 If we are mark'd to die, we are enow
To do our country loss; and if to live,
The fewer men, the greater share of honour.
William Shakespeare, *Henry V*, IV, iii

11 O when shall English men
With such acts fill a pen?
Or England breed again
Such a King Harry?
Michael Drayton, 'The Ballad of Agincourt' (1606)

---

72 HINDENBURG
Paul Ludwig von Beneckendorf und von Hindenburg, German field marshal and President of Germany (1847–1934)

---

1 Madam, I do not know exactly who won the battle, but I am quite certain who would have lost it.
Field Marshal Paul von Hindenburg, attr., on being asked if he won the Battle of Tannenberg, 1914

2 As an English general has very truly said:

'The German Army was stabbed in the back.'
Field Marshal Paul von Hindenburg,
statement to the Reichstag, 1919

3 I cannot help feeling that it were better to perish honourably than accept a disgraceful peace.
Field Marshal Paul von Hindenburg, letter to Friedrich Ebert before the Treaty of Versailles, 1919

4 Our thousand-stringed artillery began to play its battle-tune.
Field Marshal Paul von Hindenburg, *Out of My Life* (1920)

5 The soldier seeks his highest reward and his greatest joy in the consciousness of duty cheerfully done. It is character and achievement that mark out his path and measure his worth.
Field Marshal Paul von Hindenburg,
Proclamation to the German Army, 1934

6 The name Hindenburg is a terror to our enemies; it electrifies our army and our people, who have boundless faith in it.
Theobald von Bethmann-Hollweg, after Hindenburg's victory at the Battle of Tannenberg, 1914, quoted in Craig, *Germany 1866–1945* (1978)

7 The symbol of victory for the German people — Kitchener and Joffre, as it were, rolled into one.
A.J.P. Taylor, *The First World War* (1963)

---

73 HITLER
Adolf, German dictator (1889–1945)

---

1 The efficiency of the truly rational leader consists primarily in preventing the division of the attention of a people, and always concentrating it on a single enemy.
Adolf Hitler, *Mein Kampf* (1924)

2 The one means that wins the easiest victory over reason: terror and force.
Adolf Hitler, *Mein Kampf* (1924)

3 The territory on which one day our German peasants will be able to bring forth and nourish their sturdy sons will justify the blood of the sons of peasants that has to be shed today.
Adolf Hitler, *Mein Kampf* (1924)

4 There is only one right in the world and that right is one's own strength.
Adolf Hitler, speech at Nuremberg, 1928

5 Our men will be trained to become a hard breed. The German woman can be assured that the coming generations of men will be in very truth the shield and shelter of their women.
Adolf Hitler, speech at Nuremberg, 1937

6 I go the way that Providence dictates with the assurance of a sleepwalker.
Adolf Hitler, speech at Munich, 1938

7 Soldiers of the Western Front, your hour has come. The fight which begins today will determine Germany's destiny for a thousand years.
Adolf Hitler, Order to the German Army, 1940

8 In this hour I feel it to be my duty before my own conscience to appeal once more to reason and common sense in Great Britain as much as elsewhere. I consider myself in a position to make this appeal since I am not the vanquished begging favours, but the victor speaking in the name of reason. I can see no reason why this war must go on.
Adolf Hitler, speech in the Reichstag after the fall of France, 1940

9 I am convinced that 1941 will be the crucial year of a great New Order in Europe. The world shall open up for everyone.
Adolf Hitler, speech in Berlin, 1941

10 The nation has proved itself weak, and the future belongs solely to the stronger Eastern nation. Besides, those who remain after the battle are of little value; for the good have fallen.
Adolf Hitler, Order of the Day, 1945

11 I have no intention whatever of making that Austrian Corporal either Minister of Defence or Chancellor of the Reich.
Field Marshal Paul von Hindenburg, statement before Hitler's appointment as Chancellor, 1933, quoted in Wheeler-Bennett, *The Nemesis of Power* (1953)

12 I, and many others who had interviews with him, were at first impressed by his sincerity, and later realised that he was sincere only in his belief to rule the world.
Vernon Bartlett, *This Is My Life* (1937)

13 Hitler has missed the bus.
Neville Chamberlain, speech in the House of Commons after the invasion of Norway, 1940

14 Adolph Hitler is a bloodthirsty guttersnipe, a monster of wickèdness, insatiable in his lust for blood and plunder.
Winston S. Churchill, speech in the House of Commons after Hitler's invasion of the Soviet Union, 1941

15 A racing tipster who only reached Hitler's level of accuracy would not do well for his clients.
A.J.P. Taylor, *Origins of the Second World War* (1961)

---

74 HOOKER
Joseph ('Fightin' Joe), American (Union) general (1814–79)

---

1 No one will consider the day as ended until the duties it brings have been discharged.
General Joseph Hooker, Order of the Day, on assuming command in the North-West Territories, 1865

2 In the 'Fifties' when out in California at the Gold Rush, he became famous for his 'glad eye' for ladies of easy virtue, whence the Californians invented the name 'Hookers' for the type of ladies the debonair lieutenant liked so well.
George Fort Milton, *Conflict: The American Civil War* (1941)

3 (Of the Battle of Chancellorsville) 'Fighting Joe', so famous as a subordinate, bent under the strain of supreme command.
Winston S. Churchill, *A History of the English Speaking Peoples*, vol. IV (1958)

---

75 JACKSON
Thomas Jonathan ('Stonewall') American (Confederate) general (1824–63)

---

1 If I can deceive my own friends I can make certain of deceiving the enemy.
General Thomas Jackson, quoted in Henderson, *Stonewall Jackson and the American Civil War*, vol. I (1898)

2 You may appear much concerned at my attacking on Sunday. I am greatly concerned, too; but I felt it my duty to do it.
General Thomas Jackson, letter to his wife from Kernstown, Virginia, 1862

3 Let us cross the river and rest in the shade.
General Thomas Jackson, last words, 1863

4 There is Jackson, standing like a stone wall.
Brig-Gen Barnard E. Bee, at the Battle of Bull Run, 1861

5 I never saw one of Jackson's couriers approach, without expecting an order to assault the North Pole.
Richard S. Ewell, during the Valley campaign, 1862

6 You are better off than I am, for while you have lost your *left*, I have lost my *right* arm.
General Robert E. Lee, note to Jackson when he was mortally wounded at Chancellorsville, 1863

7 He was a gallant soldier, and a Christian gentleman.
General Ulysses S. Grant, during the Wilderness campaign, 1864

8 Outwardly Jackson was not a stonewall for it was not his nature to be stable and defensive, but vigorously active. He was an avalanche from an unexpected quarter. He was a thunderbolt from a clear sky. And yet he was in character and will more like a stonewall than any man I have known.
James Prior Smith, quoted in Riley, *Stonewall Jackson* (1920)

JAN SOBIEWSKI
See 78. JOHN III

---

76 JOAN OF ARC
French saint and national heroine (1412–31)

---

1 I can never see French blood without my hair standing on end.
Joan of Arc, during the Siege of Orleans, 1429

2 If you will not believe the news sent to you by God and the Maid, we will strike into

whatever place we find you, and make such great hayhay that none so great has been made in France for a thousand years.
Joan of Arc, letter to the Duke of Bedford and the English army besieging Orleans, 1429

3 All those who make war against the holy realm of France, make war against King Jesus, king of heaven and all the world.
Joan of Arc, letter to Philip the Good, Duke of Burgundy following the coronation of King Charles VII of France, 1429

4 Everything that I have done that was good I did by command of my voices.
Joan of Arc, statement during her Trial of Condemnation, 1431

5 When I have done that for which I have been sent by God, then I shall put on women's clothes.
Joan of Arc, statement to the judges during her Trial of Condemnation, 1431

6 She was all innocence, except in arms, for I saw her riding on horseback as the best of soldiers would have done, and at that the men-at-arms marvelled.
Margaret La Touroulde, deposition at Joan of Arc's Trial of Rehabilitation, 1456

7 Pity was the inspiration of Jeanne, not the pity of a woman who weeps and groans, but the magnanimous pity of a heroine who feels called to a mission and takes the sword to succour others.
C.A. Saint-Beuve, *Les Causeries du Lundi* (1851)

77 JOFFRE
Joseph Jacques Césaire, Marshal of France (1852–1931)

1 We must let ourselves be killed rather than retreat.
Marshal J.J.C. Joffre, Proclamation to the French Army before the Battle of the Marne, 1914

2 No retreat at Verdun. We fight to the end.
Marshal J.J.C. Joffre, Order to French staff officers, Chantilly, 1916

3 At the peace it would never do for France to have no army left at all.
Marshal J.J.C. Joffre, statement during the

allied conference on the use of reserves, 1916, quoted in Callwell, *Stray Recollections*, II (1923)

4 The best and largest portion of the German army was on our soil, with its line of battle jutting out a mere five days' march from the heart of France. This situation made it clear to every Frenchman that our task consisted in defeating this enemy, and driving him out of our country.
Marshal J.J.C. Joffre, *Memoirs* (1932)

5 A radical alteration in the system of command is always delicate and even critical in the midst of war.
Marshal J.J.C. Joffre, *Memoirs* (1932)

6 The only time he ever put up a fight was when we asked him to resign.
Georges Clemenceau, 1916, quoted in Thomson, *Here I Lie* (1937)

7 He went on demanding just one more attack, convinced each time that it would be decisive.
A.J.P. Taylor, *The First World War* (1963)

78 JOHN III
('Jan Sobiewski'), King of Poland (1629–96)

1 He who captures booty must be at the front.
King John III, letter to his wife after raising the Siege of Vienna, 1683, quoted in Kukulski, *Listy do Marysienki* (1973)

2 Thanks to Heaven, now the Half-Moon triumphs no longer over the Cross.
King John III, letter to his wife after raising the Siege of Vienna, 1683, quoted in Kukulski, *Listy do Marysienki* (1973)

3 They don't want to listen to me when I'm alive, so why should they obey me when I'm dead?
King John III, last words before his death, 1696

4 There is the other fool who wasted his time fighting the Turks.
Tsar Nicholas I, on viewing Sobiewski's statue in Warsaw, 1828, quoted in Davies, *God's Playground*, I (1981)

79 JOSEPH
Chief of the Nez Perce (1831–1904)

1 War can be avoided, and it ought to be avoided. I want no war. My people have always been the friend of the white man.
Chief Joseph, statement to General Oliver Howard, Wallowa Valley, 1877

2 I would have given my own life if I could have undone the killing of white men by my people.
Chief Joseph, 'An Indian's Views of Indian Affairs', *North American Review*, 1879

3 If the white man wants to live in peace with the Indian he can live in peace. There need be no trouble. Treat all men alike.
Chief Joseph, address to Congress, 1879

4 I claim a right to live on my land, and accord you the privilege to live on yours.
Chief Joseph, quoted in Howard, *War Chief Joseph* (1941)

80 JULIUS CAESAR
Roman general and dictator (100–44 BC)

1 *Iacta alea est.* The die is cast.
Julius Caesar, at the crossing of the Rubicon, quoted in Suetonius, *Divus Julius*, 32

2 *Veni, vidi, vici.* I came, I saw, I conquered.
Julius Caesar, quoted in Suetonius, *Divus Julius*, 37, 2

3 *Cur denique fortunam periclitaretur? Praesertum quum non minus esset imperatoris, consilio superare, quam gladio.* Why stake your fortunes on the risk of battle? Especially as a victory by strategy is as much a part of good generalship as a victory by the sword.
Julius Caesar, *De Bello Civili*, I, 72

4 *Fortuna, quae plurimum potest, quum in reliquis rebus, tum praecipue in bello, parvis momentis magnas rerum communitationes efficit.* All-powerful fortune, in war above all things, produces momentous changes from very small beginnings.
Julius Caesar, *De Bello Civili*, III, 68

5 *Gallia est omnis divisa in partes tres.* The whole of Gaul is divided into three parts.
Julius Caesar, *De Bello Gallico*, I, i

6 In war trivial causes produce momentous events.
Julius Caesar, *De Bello Gallico*, I

7 *Nemo est tam fortis, qui rei novitate perturbetur.* No one is so brave as not to be disconcerted by unforeseen circumstances.
Julius Caesar, *De Bello Gallico*, VI, 39

8 As he was valiant, I honour him: but, as he was ambitious, I slew him. (Brutus)
William Shakespeare, *Julius Caesar*, III, ii

9 He was my friend, faithful and just to me. (Mark Antony)
William Shakespeare, *Julius Caesar*, III, ii

10 What millions died that Caesar might be great?
Thomas Campbell, *The Pleasures of Hope* (1799)

81 KITCHENER
Horatio Herbert, Earl Kitchener of Khartoum, British field marshal (1850–1916)

1 If you do not do something in the way of helping me in the entertaining I shall be driven into some rash alliance of a matrimonial nature, so you have your warning and better look out.
Field Marshal Earl Kitchener, letter to his sister Millie, 1892

2 I am afraid I rather disgust the old red-tape heads of departments. They are very polite, and after a bit present me with a volume of their printed regulations generally dated about 1870 and intended for Aldershot manoeuvres, and are quite hurt when I do not agree to follow their printed rot.
Field Marshal Earl Kitchener, letter to Pandeli Ralli during the Boer War, 1900

3 War means risks, and you cannot play the game and always win; and the sooner those in authority realise this, the better.
Field Marshal Earl Kitchener, letter to Lady Cranborne, 1900

4 (Of the Boer War) We are now carrying on

the war to be able to put 2-300 Dutchmen in prison at the end of it. It seems to me to be absurd, and I wonder the Chancellor of the Exchequer did not have a fit.
Field Marshal Earl Kitchener, letter to St John Brodrick, 1902

5 Though you may have to toil as before, you can never quite be the same again. You have tasted the salt of life, and its savour will never leave you.
Field Marshal Earl Kitchener, addressing his army in Johannesburg at the conclusion of the Boer War, 1902

6 In the old days I suppose I should have called him out and shot him like a dog for his grossly insulting letter.
Field Marshal Earl Kitchener, letter to Lady Salisbury describing a disagreement with Lord Curzon, 1905

7 I am doing all that I can and yet I feel that I am still leaving much undone.
Field Marshal Earl Kitchener, to Lord Derby, 1915, quoted in Royle, *The Kitchener Enigma* (1985)

8 Rightly or wrongly, probably wrongly, the people believe in me. It is not therefore me that the politicians are afraid of, but what the people would say to them if I were to go.
Field Marshal Earl Kitchener, letter to Field Marshal Earl Haig, 1916

9 (Of the Western Front) We had to make war as we must, not as we would like to.
Field Marshal Earl Kitchener, address to War Cabinet, 1916

10 The man I have always placed my hopes upon, Major Kitchener, RE, who is one of the very few superior officers with a hard constitution, combined with untiring energy.
General Charles George Gordon, *Journals* (1885)

11 He really feels nice things, but to put tongue to them, except in very intimate society, he would rather die.
Brig-Gen Frank Maxwell, VC, letter to his father, 1901

12 Kitchener is a strong man, but all he is doing is to paralyse me.
Alfred Milner, quoted in Headlam, *The Milner Papers,* vol. II (1931–3)

13 He stands aloof and alone, a molten mass of devouring energy and burning ambition, without anybody to control or guide it in the right direction.
Lord Curzon, quoted in Dilks, *Curzon in India,* vol. I (1970)

14 There is a sort of bracing north wind of resolution and strenuousness about him, and a gentle spot about a woman which a woman is always quick to find.
Lady Curzon, quoted in Nicolson, *Mary Curzon* (1977)

15 Your Country Needs You.
First World War recruiting poster

16 Lord K is playing Hell with its lid off at the War Office — What the papers call 'standing no nonsense' but which often means listening to no nonsense.
Lady Jean Hamilton, Diary, 1914

17 I knew the solid advantages of that wonderful name and personality, with the power to move people and inspire them to patriotic effort.
Sir Hedley Le Bas, on Kitchener's support for the Parliamentary Recruiting Committee in 1914, quoted in *The Kitchener Memorial Book* (1917)

18 Kitchener is not attractive. None of the men who served with him were attracted to him. It is the coarseness of his fibre, which appears in his face to a marked degree. The eyes are good — but the jaw and skin are those of a rough private.
Lord Esher, *Journals,* vol. I (1934)

19 If he was not a great man, at least he was a great poster.
Margot Asquith, *Autobiography* (1936)

82 LAFAYETTE
Marie Joseph Paul Roch Yves Gilbert Motier, Marquis de Lafayette, French general (1757–1834)

1 I hope to return more worthy of all who might be kind enough to regret my departure.
Marquis de Lafayette, letter to his father on departing for America to join the American rebels, 1777

2 The fortune of America is closely bound up with the fortune of humanity: she will become the safe and respected refuge of

virtue, honesty, tolerance, equality and of a peaceful liberty.
Marquis de Lafayette, letter to his wife from Charleston, South Carolina, 1777

3 To act in concert with a great man is the first of blessings.
Marquis de Lafayette, letter to General George Washington, 1778

4 For a nation to love liberty, it is enough that she understands it; for her to be free, it is enough that she wishes it.
Marquis de Lafayette, address to the National Assembly, Paris, 1789

5 America has joined forces with the Allied Powers, and what we have in blood and treasure is yours. With loving pride we drape the colours in tribute of respect to this citizen of your great republic. And here and now in the presence of the illustrious dead we pledge our hearts and our honour in carrying this war to a successful issue. Lafayette, we are here!
Colonel Charles E. Stanton, address at the tomb of Lafayette, Paris, 1917

83 LAWRENCE
**Thomas Edward, ('Lawrence of Arabia')
British guerrilla leader (1888–1935)**

1 If you can wear Arab kit when with the tribes you will acquire their trust and intimacy to a degree impossible in uniform.
T.E. Lawrence, *Twenty-Seven Articles* (1917)

2 It's the most amateurish, Buffalo-Billy sort of performance, and the only people who do it well are the Bedouin.
T.E. Lawrence, letter to Major W.F. Stirling after ambushing a Turkish train, 1917

3 The Army is muck, stink and desolate abomination.
T.E. Lawrence, letter to Eric Kennington, 1923

4 In a madness born of the horror of Tafas we killed and killed, even blowing in the heads of the fallen and of the animals; as though their death and running blood could slake our agony.
T.E. Lawrence, *The Seven Pillars of Wisdom* (1926)

5 The Arab inspiration was our main tool in winning the Eastern War.
T.E. Lawrence, *The Seven Pillars of Wisdom* (1926)

6 We should have more bright breasts in the Army if each man was able, without witness, to write his own despatch.
T.E. Lawrence, *The Seven Pillars of Wisdom* (1926)

7 He had a genius for backing into the limelight.
Lowell Thomas, *With Lawrence in Arabia* (1925)

8 He made so many mysteries; told one person one set of facts, another, another. If most compared notes most would be known, but not all.
Robert Graves, letter to Basil Liddell Hart, 1931

9 He over-awed his comrades, frightened sergeants, made squadron leaders feel uncomfortable and patronised the Air Council.
W.B. Tawse, letter to the *Evening Standard*, 1973

10 (Of the film *Lawrence of Arabia*) They only got two things right, the camels and the sand.
Lowell Thomas, quoted in his obituary, *The Times*, 1981

84 LEE
Robert Edward, American (Confederate) general (1807–70)

1 A soldier has a hard life, and but little consideration.
General Robert E. Lee, letter to his wife, 1855

2 What a cruel thing is war.
General Robert E. Lee, letter to his wife, 1862

3 It is well that war is so terrible — we would grow too fond of it.
General Robert E. Lee, to James Longstreet, at Fredericksburg, 1862

4 Duty is the sublimest word in our language.

Do your duty in all things. You cannot do more. You should never do less.
General Robert E. Lee, inscription beneath his bust, Hall of Fame for Great Americans, New York

5 Lee is the only man I know whom I would follow blindfold.
General Thomas Jackson, letter, 1862

6 Posterity will rank Lee above Wellington or Napoleon, before Saxe or Turenne, above Marlborough or Frederick, before Alexander or Caesar.
Anon., obituary in the *Montreal Telegraph*, 1870

7 If I were on my death bed, and the President should tell me that a great battle was to be fought for the liberty or slavery of my country, and asked my judgement as to the ability of a commander, I would say with my dying breath, let it be Robert E. Lee.
General Winfield Scott, quoted in Davis, *They Called Him Stonewall* (1954)

---

85 LESLIE
Alexander, Earl of Leven, Scottish general (1580–1661)

---

1 I wronged not my conscience in doeing any thing I was commanded to doe by these whom I served.
General Alexander Leslie, quoted in Turner, *Memoirs* (1829)

2 Such was the wisdom and authoritie of that old, little, crokked souldier, that all, with ane incredible submission, from the beginning to the end, gave over themselves to be guided by him.
Robert Baillie, *Letters* (1841–2)

3 He admirably combined authority and tact so as to get himself obeyed without causing offence.
David Stevenson, *The Scottish Revolution, 1637–1644*, (1973)

---

86 LIDDELL HART
Captain Sir Basil Henry, British military theorist (1895–1970)

---

1 The nation which is the quickest to realise that the period on which we are entering is the tank era will win the next war.
Captain Sir Basil Liddell Hart, letter to Maj-Gen J.F.C. Fuller, 1922

2 New weapons would seem to be regarded merely as an additional tap through which the bath of blood can be filled all the sooner.
Captain Sir Basil Liddell Hart, *Paris, Or the Future of War* (1925)

3 Once appreciate that tanks are not an extra arm or a mere aid to infantry but the modern form of heavy cavalry and their true military use is obvious — to be concentrated and used in as large masses as possible for a decisive blow against the Achilles heel of the enemy army, the communications and command centres which form its nerve system.
Captain Sir Basil Liddell Hart, writing in the *Westminster Gazette*, 1925

4 Loyalty to truth coincides with true loyalty to the Army in compelling a new honesty in examining and facing the facts of history.
Captain Sir Basil Liddell Hart, *The Ghost of Napoleon* (1933)

5 Since Marlborough the British Army has rarely shone in the offensive, not through want of courage, but from lack of aptitude.
Captain Sir Basil Liddell Hart, *Europe in Arms* (1937)

6 The British temperament is unsuited to offensive operations.
Captain Sir Basil Liddell Hart, letter to *The Times*, 1937

7 The advent of 'automatic warfare' should make plain the absurdity of warfare as a means of deciding nations' claims to superiority.
Captain Sir Basil Liddell Hart, *The Revolution in Warfare* (1946)

8 The Captain speaks of battles and wars and deadly weapons with a sort of cold academic calm, as if he were giving a lesson in Latin.
Raphael Bashan, interview with Captain Sir Basil Liddell Hart in *Ma'ariv*, 1960

9 His influence was most remarkable and affected, in my opinion, the whole outlook and training of the army.
Field Marshal Sir Claude Auchinleck, letter to Nicholas Eadon, 1967

87 LOUIS XIV
King of France (1643–1715)

1   *L'état c'est moi.* I am the state.
King Louis XIV, attr. remark before the
Parlement de Paris, 1655, quoted in
Dulaure, *Histoire de Paris* (1834)

2   *Il n'y a plus de Pyrénées.* The Pyrenees have
ceased to exist.
King Louis XIV, on the accession of his
grandson to the throne of Spain, 1700

3   How could God do this to me after all I have
done for him?
King Louis XIV, on hearing of the French
defeat at the Battle of Blenheim, 1704

4   Has God forgotten all I have done for him?
King Louis XIV, on hearing of the French
defeat at the Battle of Malplaquet, 1709

5   I have loved war too well.
King Louis XIV, last words, 1715

88 LUDENDORFF
Erich von, German general (1865–1937)

1   A fortress that surrenders without having
defended itself to the last is dishonoured.
General Erich von Ludendorff, advice to
Kaiser Wilhelm II during surrender
negotiations, 1918, quoted in Tschuppik,
Ludendorff: The Tragedy of a Specialist
(1932)

2   I should be no better than a gambler if, in
view of the gravity of the situation, I did not
insist upon ending the war by asking for an
immediate armistice.
General Erich von Ludendorff, address to
his staff officers, 1918

3   Germany, keep your army powerful and
tightly in your hand and look to the horizon
again. And your recovery will come.
General Erich von Ludendorff, letter to
Wilhelm Breucker, 1919

4   August 8th (1918) was the black day of the
German Army in the history of this war.
General Erich von Ludendorff, *My War
Memoirs* (1919)

5   The offensive is the most effective means of
making war; it alone is decisive.
General Erich von Ludendorff, *My War
Memoirs* (1919)

6   War is the highest expression of the racial
will to live.
General Erich von Ludendorff, *My War
Memoirs* (1919)

7   The soldier has no personal ends to pursue.
He goes where he's sent.
General Erich von Ludendorff, quoted in
Margarethe Ludendorff, *My Married Life
with Ludendorff* (1930)

8   We do not want a union of the Rhine
resulting from French favours, nor a state
under the influence of Marxist–Jewish or
supra-national powers, but a Germany that
would only belong to the Germans, in which
nothing rules but German will, German
honour and German strength.
General Erich von Ludendorff, *Auf dem
Weg zur Feldherrnhalle* (1937)

9   Is it not wonderful for a regimental
commander when his young lieutenants call
him father?
Wilhelm Breucker, speaking of Ludendorff
in 1914, quoted in Margarethe Ludendorff,
My Married Life with Ludendorff (1930)

89 MACARTHUR
Douglas, American general (1880–1964)

1   A good soldier, whether he leads a platoon
or an army, is expected to look backward as
well as forward, but he must think only
forward.
General Douglas MacArthur, address to
West Point graduates, 1933

2   I have come through and I will return.
General Douglas MacArthur, address to
Australians after the fall of the Philippines,
1942

3   We shall win or we shall die, and to this end
I pledge you the full resources of the mighty
power of my country and all the blood of my
countrymen.
General Douglas MacArthur, speech in
Melbourne, 1942

4   I see that the old flagpole still stands. Have

your troops hoist the colours to its peak and let no enemy haul them down.
General Douglas MacArthur, Order to his troops, Philippines, 1945

5 However horrible the incidents of war may be, the soldier who is called upon to offer and to give his life for his country is the noblest development of mankind.
General Douglas MacArthur, speech to veterans, Washington, 1947

6 Like the old soldier of the ballad, I now close my military career and just fade away.
General Douglas MacArthur, address to Congress, 1951

7 War's very object is victory, not prolonged indecision.
General Douglas MacArthur, address to Congress, 1951

8 It is fatal to enter any war without the will to win it.
General Douglas MacArthur, speech at Republican National Convention, 1952

9 Your mission remains fixed, determined, inviolable — it is to win wars. Everything else in your professional career is but corollary to this vital dedication.
General Douglas MacArthur, address to cadets at West Point, 1962

10 Everything about Douglas MacArthur is on the grand scale; his virtues and triumphs and shortcomings.
Lt-Gen George H. Brett, in *True*, 1947

11 The best and the worst things you hear about him are quite true.
General Sir Thomas Blamey, quoted in Hetherington, *Blamey* (1954)

---

90 MACHIAVELLI
Niccolo, Italian statesman and political philosopher (1469–1527)

---

1 A necessary war is a just war.
Niccolo Machiavelli, *The Prince* (1513)

2 There cannot be good laws where there are not good arms.
Niccolo Machiavelli, *The Prince* (1513)

3 War should be the only study of a prince.
Niccolo Machiavelli, *The Prince* (1513)

4 When princes think more of luxury than of arms, they lose their state.
Niccolo Machiavelli, *The Prince* (1513)

5 Whoever conquers a free town and does not demolish it commits a great error and may expect to be ruined himself.
Niccolo Machiavelli, *The Prince* (1513)

6 Impetuosity and audacity often achieve what ordinary means fail to achieve.
Niccolo Machiavelli, *Discorsi* (1531)

7 Men rise from one ambition to another; first they seek to secure themselves from attack, and then they attack others.
Niccolo Machiavelli, *Discorsi* (1531)

8 One should never risk one's whole fortune unless supported by one's entire forces.
Niccolo Machiavelli, *Discorsi* (1531)

9 Though fraud in other activities be detestable, in the management of war it is laudable and glorious, and he who overcomes an army by fraud is as much to be praised as he who does so by force.
Niccolo Machiavelli, *Discorsi* (1531)

10 We are much beholden to Machiavel and others, that write what men do, and not what they ought to do.
Francis Bacon, *The Advancement of Learning*, II (1605)

11 Machiavelli was not an evil genius, nor a cowardly and miserable writer, he is nothing but the fact. And he is not merely the Italian fact, but the fact of the sixteenth century.
Victor Hugo, *Les Misérables* (1862)

---

91 MACMAHON
Marie Edmé Patrice Maurice, Comte de MacMahon, Marshal of France and President of France (1808–93)

---

1 Here I am, here I stay.
Marshal Maurice MacMahon, on capturing Malakoff Fort, 1855

2 If they attack us, so much the better; we shall be able, no doubt, to fling them into the Meuse.
Marshal Maurice MacMahon, Order to General Lebrun before the defeat of the French army at the Battle of Sedan, 1870

92 MAO TSE-TUNG
Chinese soldier and statesman
(1893–1976)

1 Our principle is that the Party commands
the gun and the gun will never be allowed to
command the Party.
Mao Tse-Tung, 'Problems of War and
Strategy', *Selected Works*, vol. II (1954)

2 All reactionaries are paper tigers.
Mao Tse-Tung, *Quotations* (1966)

3 In order to get rid of the gun, it is necessary
to take up the gun.
Mao Tse-Tung, *Quotations* (1966)

4 The people, and the people alone, are the
motive force in the making of world history.
Mao Tse-Tung, *Quotations* (1966)

5 Politics is war without bloodshed, war is
politics with bloodshed.
Mao Tse-Tung, *Quotations* (1966)

6 Political power grows out of the barrel of a
gun.
Mao Tse-Tung, *Quotations* (1966)

7 We Communists are like seeds and the
people are like the soil. Wherever we go, we
must unite with the people, take root and
blossom among them.
Mao Tse-Tung, *Quotations* (1966)

93 MARLBOROUGH
John Churchill, Duke of Marlborough,
British soldier (1650–1722)

1 If the occasion for a fight offers you can rest
assured that I shall not neglect it.
Duke of Marlborough, letter to Prince
Eugene of Savoy, 1706

2 If the siege of Toulon goes properly I shall
be cured of all diseases except old age.
Duke of Marlborough, letter to his wife,
1707

3 I hope I have given such a blow to their foot
that they will not be able to fight any more
this year.
Duke of Marlborough, letter to the Earl of
Godolphin after the Battle of Oudenarde,
1708

4 It is now in our power to have what peace
we please, and I may be pretty well assured
of never being in another battle.
Duke of Marlborough, letter to his wife
after the Battle of Malplaquet, 1709

5 That was once a man.
Duke of Marlborough, on contemplating
his portrait in old age, quoted in Churchill,
Marlborough, His Life and Times, vol. IV
(1938)

6 In peaceful thought the field of death
survey'd
To fainting squadrons sent the timely aid,
Inspir'd, repuls'd battalions to engage,
And taught the doubtful battle where to
rage.
Joseph Addison, *The Campaign* (1704)

7 The Duke returned from the wars today and
did pleasure me in his top-boots.
Sarah, Duke of Marlborough, attr. in
various forms, 1711

8 The known world could not produce a man
of more humanity.
Corporal Matthew Bishop, *Life and
Adventures* (1744)

9 He never fought a battle which he did not
gain, nor laid siege to a town which he did
not take.
Captain Robert Parker, *Memoirs* (1747)

10 There was a certain personal quality in
Marlborough's type of command which is
of especial interest, because it was national;
I mean the solidity of his judgement.
Hilaire Belloc, *The Tactics and Strategy of
the Great Duke of Marlborough* (1933)

11 Marlborough was perhaps the only great
general to whom geniality was always
natural.
Field Marshal Earl Wavell, *Generals and
Generalship* (1941)

94 MARSHALL
George Catlett, American general
(1880–1959)

1 It was not the ordeal of personal combat
that seemed to prove the greatest strain in
the last war. It was the endurance for days at
a time of severe artillery bombardment by

shells of heavy calibre that proved the fortitude of the troops.
General George Marshall, Diary, 1917

2 Notwithstanding the entry of Japan into the war, our view is that Germany is still the prime enemy and her defeat is the key to victory. Once Germany is defeated, the collapse of Italy and the defeat of Japan must follow.
General George Marshall, memorandum at Arcadia Conference, Washington, 1941, quoted in Sherwood, *Roosevelt and Hopkins* (1948)

3 The single objective should be quick and complete victory.
General George Marshall, signal to General Eisenhower before the Rhine crossing, 1945

4 As architect and builder of the finest and most powerful Army in American history, your name will be honoured among those of the greatest soldiers of our own or any other country.
Field Marshal Viscount Alanbrooke, letter to General Marshall on his retirement as Chief of Staff, 1945

5 It has not fallen to your lot to command great armies. You have had to create them, organise them, and inspire them.
Winston S. Churchill, letter to General Marshall after the defeat of Nazi Germany, 1945

4 To leave a country defenceless would be the greatest crime of its government.
Helmuth, Graf von Moltke, speech in the Reichstag, 1874

5 The fate of every nation lies in its own strength.
Helmuth, Graf von Moltke, speech in the Reichstag, 1880

6 No plan of operations reaches with any certainty beyond the first encounter with the enemy's main force.
Helmuth, Graf von Moltke, *Kriegsgeschichtiche Einzelschriften* (1880)

7 War is part of God's world order.
Helmuth, Graf von Moltke, letter to J.K. Bluntschli, 1880

8 It is no longer the ambition of princes; it is the moods of the people, the discomfort in the face of interior conditions, the doings of parties, particularly of their leaders, which endanger peace. The great battles of recent history have started against the wish and will of the governors.
Helmuth, Graf von Moltke, *Gesammelte Schriften*, III (1888)

9 War was his trade.
Count Otto von Bismarck, *Gedanken und Erinnerungen* (1891)

95 MOLTKE
Helmuth Karl Berhard, Graf von Moltke, Prussian field marshal (1800–91)

1 The character of modern military leadership in war is shown in the pursuit of a large and swift decision.
Helmuth, Graf von Moltke, motto adopted after the Luxembourg crisis, 1867

2 To remain separated as long as possible while operating and to be concentrated in good time for the decisive battle, that is the task of the leader of large masses of troops.
Helmuth, Graf von Moltke, Instructions for Superior Commanders, 1869

3 The Prussian schoolmaster won the battle of Sadowa.
Helmuth, Graf von Moltke, speech in the Reichstag, 1874

96 MONASH
Sir John, Australian general (1865–1931)

1 The main thing is always to have a plan; if it is not the best plan, it is at least better than no plan at all.
General Sir John Monash, letter of 1918

2 I hate the business of war — the horror, the waste, the destruction, the inefficiency.
General Sir John Monash, *War Letters* (1935)

3 The Australian soldier had the political sense highly developed, and was always a keen critic of the way in which his battalion or battery was 'run', and of the policies which guided his destinies from day to day.
General Sir John Monash, *The Australian Victories in France in 1918* (1936)

4 I don't believe he thought about anything during the War except winning the war.
Sir Keith Officer, letter to John Terraine, 1958, quoted in Terraine, *The Smoke and the Fire* (1980)

5 The only general of creative originality produced by the First World War.
A.J.P. Taylor, *The First World War* (1963)

97 MONCK
George, Duke of Albemarle, English soldier and sailor (1608–70)

1 Obedience is my great principle and I have always, and ever shall, reverence the Parliament's resolutions in civil things as infallible as sacred.
General George Monck, appeal to the Speaker of the House of Commons, 1659

2 If the army will stick by me, I will stick by them.
General George Monck, Order committing his forces to the side of the Rump of the Long Parliament, 1659

3 Before you undertake a War cast an impartial eye upon the cause. If it be just, prepare your Army and let them all know they fight for God.
General George Monck, *Observations Upon Military and Political Affairs* (1671)

4 The profession of a soldier is allowed to be lawful by the Word of God and so famous and honourable among men, that emperors and kings do account it a great honour to be of the profession.
General George Monck, *Observations Upon Military and Political Affairs* (1671)

5 Reading and Discourse are requisite to make a Souldier perfect in the Art Military, how great soever his practical knowledge may be.
General George Monck, *Observations Upon Military and Political Affairs* (1671)

6 It is glorious enough to his memory that he was instrumental in bringing mighty things to pass which he had neither wisdom to foresee nor courage to attempt nor understanding to continue.
Earl of Clarendon, *History of the Rebellion* (1704)

7 Monk was one of those Englishmen who understand to perfection the use of time and circumstance. It is a type which has thriven in our island.
Winston S. Churchill, *A History of the English Speaking Peoples*, vol. II (1956)

98 MONTCALM
Louis Joseph, Marquis de Montcalm, French soldier (1712–59)

1 (Of the Heights of Abraham) I swear to you that a hundred men posted there would stop their whole army.
Marquis de Montcalm, letter to Marquis de Vaudreuil-Cavagnal, 1759

2 Oh, when shall we get out of this country! I think that I would give half that I have to get home.
Marquis de Montcalm, letter to his wife, 1759

3 Oh that I had served my Lord as faithfully as I have served my King.
Marquis de Montcalm, quoted in Parkman, *Montcalm and Wolfe* (1884)

4 War is the grave of the Montcalms.
Marquis de Montcalm, quoted in Parkman, *Montcalm and Wolfe* (1884)

5 Destiny, in robbing him of victory, compensated him with a glorious death.
Lord Aylmer, inscription on Montcalm's Monument, Plains of Abraham, 1831

99 MONTECUCCOLI
Raimundo, Italian soldier (1609–80)

1 For war you need three things: 1. Money, 2. Money, 3. Money.
General Raimundo Montecuccoli, attr., quoted in Keegan and Wheatcroft, *Who's Who in Military History* (1976)

100 MONTGOMERY
Bernard Law, Viscount Montgomery of Alamein, British field marshal (1887–1976)

1 We will hit the enemy for six right out of North Africa.
Field Marshal Viscount Montgomery, message to his troops, 1942

2   You can't run a military operation with a
    committee of staff officers in command. It
    will be nonsense.
    Field Marshal Viscount Montgomery,
    1944, quoted in de Guingand, *Operation
    Victory* (1947)

3   The point to understand is that if we had
    run the show properly, the war could have
    been finished by Christmas 1944. The
    blame for this must rest with the Americans.
    Field Marshal Viscount Montgomery,
    'Notes on the Campaign in North-West
    Europe', War Diary, 1945

4   War grows directly out of things which
    individuals, statesmen and nations *do* or fail
    to do.
    Field Marshal Viscount Montgomery,
    address to British and Commonwealth army
    officers, Camberley, 1948

5   The British soldier is second to none in the
    communities of fighting men.
    Field Marshal Viscount Montgomery,
    Memoirs (1958)

6   The British soldier responds to leadership in
    a most remarkable way; and once you have
    won his heart, he will follow you anywhere.
    Field Marshal Viscount Montgomery,
    Memoirs (1958)

7   Leadership which is evil, while it may
    temporarily succeed, always carries within
    itself the seeds of its own destruction.
    Field Marshal Viscount Montgomery,
    Memoirs (1958)

8   No leader, however great, can long continue
    unless he wins victories. The battle decides
    all.
    Field Marshal Viscount Montgomery,
    Memoirs (1958)

9   One of the big lessons I learned from the
    campaign in Africa was the need to decide
    what you want to do, and then to *do* it.
    Field Marshal Viscount Montgomery,
    Memoirs (1958)

10 I've got to go to meet God — and explain all
    those men I killed at Alamein.
    Field Marshal Viscount Montgomery,
    1976, quoted in Hamilton, *Monty: The
    Field Marshal* (1986)

11 He revealed no trace of ordinary human
    frailties or foibles.
    Chester Wilmot, *The Struggle for Europe*
    (1952)

12 In defeat, unbeatable; in victory,
    unbearable.
    Winston S. Churchill, quoted in Marsh,
    Ambrosia and Small Beer (1964)

13 He was not a convivial man but he was a
    ready man, with a sharp eye and a sharp
    tongue, and he certainly made himself
    known.
    General Sir John Hackett, *The Profession of
    Arms* (1983)

14 The record of his self-preparation and the
    training of the officers and men beneath him
    probably has no parallel in the history of
    British arms since the days of Wellington
    and Nelson.
    Nigel Hamilton, *Monty: Master of the
    Battlefield* (1983)

15 For all Montgomery's caution in battle, the
    passion for 'tidiness' that more than once
    denied him all-embracing victories, this
    essentially cold, insensitive man was
    devoted to winning.
    Max Hastings, *Overlord* (1984)

---

101 MONTROSE
   James Graham, Marquis of Montrose,
   Royalist general (1612–50)

---

1   Though Caesar's paragon I cannot be,
    Yet shall I soar in thought as high to be.
    James Graham, Marquis of Montrose,
    epigram in his copy of Caesar's
    Commentaries, c. 1629

2   He either fears his Fate too much,
    Or his Deserts are small,
    That puts it not unto the Touch,
    To win or lose it all.
    James Graham, Marquis of Montrose,
    'Montrose to his Mistress' (*c.* 1640)

3   The commands of my sovereign were to
    defend his safety in his deep distress against
    wicked rebels in arms against him. It was my
    duty to obey.
    James Graham, Marquis of Montrose,
    address to the people of Edinburgh before
    his execution, 1650

4   That was the true gentleman,
    Who came of line not humble.

Good was the flushing of his cheek
When drawing up to battle.
Iain Lom, 'Lament for Montrose', trs.
Nicholson

5 Perhaps the most brilliant natural military
genius disclosed by the Civil War.
Sir John Fortescue, *A History of the British
Army* (1899–1912)

102 MOORE
Sir John, British general (1761–1809)

1 To be killed is certainly in some degree more
honourable, and certainly more pleasing, to
a soldier.
General Sir John Moore, letter to his father
from America, 1782

2 It is melancholy to be obliged to act against
one's countrymen. I hope sincerely it will
never be our case; but, as soldiers, we
engaged not only to fight the foreigner; but
also to support the Government and laws,
which have long been in use and framed by
wiser men than we are.
General Sir John Moore, address to his Irish
troops, Cork, 1798

3 The discipline of the mind is as requisite as
that of the body to make a good soldier.
General Sir John Moore, letter to Robert
Brownrigg, 1804

4 How they came to pitch upon me I cannot
say, for they have given sufficient proof of
not being partial to me.
General Sir John Moore, Diary, on being
made Commander-in-Chief in Spain, 1808

5 They talked of going into Spain as if going
into Hyde Park.
General Sir John Moore, of his army before
the retreat to Corunna, in a letter to Lord
William Bentinck, 1808

6 I feel myself so strong. I fear I shall be a long
time in dying.
General Sir John Moore, on being mortally
wounded at the Battle of Corunna, 1809

7 I wish Sir John had united something of the
Christian with the hero in his death.
Jane Austen, letter of 1809

8 Not a drum was heard, nor a funeral note,
As his corpse to the rampart we hurried;

Not a soldier discharged his farewell shot
O'er the grave where our hero we buried.
Charles Wolfe, 'The Burial of Sir John
Moore After Corunna', *Newry Telegraph*
(1817)

9 If glory be a distinction, for such a man
death is not a leveller.
Maj-Gen Sir William Napier, *Peninsular
War*, vol. I (1886)

103 MOUNTBATTEN
Louis Francis Albert Victor Nicolas, Earl Mountbatten of Burma, British admiral and military leader (1889–1979)

1 I want to make it clear to all of you that I
shall never give the order 'abandon ship',
the only way you can leave the ship is if she
sinks beneath your feet.
Admiral Earl Mountbatten, Order to the
crew of HMS *Kelly*, 1940

2 (Of the Japanese generals) If I had my way
I'd shoot about twenty of them — you've
got to do something to satisfy the bloodlust.
Then I'd officially kick about 200 or 300 of
them in the arse in front of all the rest, and
I'd let them go back to their countries with
reprimands.
Admiral Earl Mountbatten, quoted in Diary
of General H.H. Arnold, 1945

3 Normally I am not a vindictive person, but I
cannot help feeling that unless we are tough
with all the Japanese leaders they will be
able to build themselves up eventually for
another war.
Admiral Earl Mountbatten, letter to
General Douglas MacArthur, 1945

4 Youthful, buoyant, picturesque, with a
reputation for gallantry known everywhere,
he talked to the British soldier with
irresistible frankness and charm.
Field Marshal Viscount Slim, *Defeat into
Victory* (1956)

104 MUSSOLINI
Benito, Italian dictator (1883–1945)

1 For my part I prefer fifty thousand rifles to
fifty thousand votes.
Benito Mussolini, remark, 1921

2   My ambition is this: I want to make the people of Italy strong, prosperous and free.
Benito Mussolini, speech to Italian Senate, 1923

3   Italy wants peace and quiet, work and calm. I will give these things with love if possible, and with force if necessary.
Benito Mussolini, speech in Chamber of Deputies, 1925

4   I am proud to have reddened the road to Trieste with my own blood in the fulfilment of my dangerous duty.
Benito Mussolini, *My Autobiography* (1928)

5   War alone brings up to their highest tension all human energies and imposes the stamp of nobility upon the peoples who have the courage to make it.
Benito Mussolini, 'The Political and Social Doctrine of Fascism', in *Enciclopedia Italiana* (1932)

6   The Germans should allow themselves to be guided by me if they wish to avoid unpardonable blunders. In politics it is undeniable that I am more intelligent than Hitler.
Benito Mussolini, remark, 1934

7   Thirty centuries of history enable us to look with majestic pity at certain doctrines taught on the other side of the Alps by the descendants of people who were wholly illiterate in the days when Rome boasted a Caesar, a Virgil and an Augustus.
Benito Mussolini, speech at Bari, 1934

8   When Fascism has a friend, it will march with that friend to the last.
Benito Mussolini, speech in Berlin, 1937

9   Fate! Statesmen only talk of fate when they have blundered.
Benito Mussolini, remark, 1942

10 Have no fear for ultimate victory. It is certain that your sacrifices will be rewarded. That is as true as it is true that God is just and that Italy is immortal.
Benito Mussolini, speech at Rome, 1943

11 This whipped jackal is frisking by the side of the German tiger.
Winston S. Churchill, speech in the House of Commons, 1941

12 Italy's pinchbeck Caesar.
Winston S. Churchill, speech at the Guildhall, London, 1943

105 NAPOLEON I
Emperor of France (1769–1821)

1   Soldiers, consider that from the summit of these pyramids, forty centuries look down upon you.
Napoleon I, address to the French Army before the Battle of the Pyramids, 1798

2   I am the successor, not of Louis XIV, but of Charlemagne.
Napoleon I, on his coronation by Pope Pius VII, 1804

3   From the sublime to the ridiculous there is only one step.
Napoleon I, after the French Army's retreat from Moscow, 1812

4   Treaties are observed as long as they are in harmony with interests.
Napoleon I, *Maxims* (1804–15)

5   Who saves his country violates no law.
Napoleon I, *Maxims* (1804–15)

6   Soldiers are made on purpose to be killed.
Napoleon I, letter to General Gaspard Gourgand, 1818

7   England is a nation of shopkeepers.
Napoleon I, quoted in O'Meara, *Napoleon in Exile* (1822)

8   Had I succeeded, I should have died with the reputation of the greatest man that ever lived.
Napoleon I, quoted in O'Meara, *Napoleon in Exile* (1822)

9   I love a brave soldier who has undergone that baptism of fire, whatever nation he belongs to.
Napoleon I, quoted in O'Meara, *Napoleon in Exile* (1822)

10 In order to have good soldiers, a nation must always be at war.
Napoleon I, quoted in O'Meara, *Napoleon in Exile* (1822)

11 At the head of an army, nothing is more becoming than simplicity.
Napoleon I, *Political Aphorisms* (1848)

12 Two Great Powers, such as France and England, if they could agree, might govern the world.
Napoleon I, *Political Aphorisms* (1848)

13 Napoleon is a torrent which as yet we are unable to stem. Moscow will be the sponge that will suck him dry.
Marshal Mikhail Kutznov, address to his troops, 1812

14 Crushed was Napoleon by the northern Thor,
Who knocked his army down with an icy hammer.
Lord Byron, *Beppo* (1818)

15 Although too much of a soldier among sovereigns, no one could claim with better right to be a sovereign among soldiers.
Sir Walter Scott, *Life of Napoleon* (1827)

16 I used to say of him that his presence on the field made the difference of forty thousand men.
Duke of Wellington, letter to Lord Stanhope, 1831

17 Napoleon will live when Paris is in ruins; his deeds will survive the dome of the Invalides — no man can show the tomb of Alexander!
Sir Archibald Alison, *The History of Europe* (1842)

18 Napoleon attempted the impossible, which is beyond even genius.
Charles Ardant du Picq, *Battle Studies* (1870)

---

106 NEY
Michel, Duc d'Elchingen, Marshal of France (1769–1815)

---

1 The wine is poured; and we must drink it.
Marshal Michel Ney, Order to advance before the Battle of Jena, 1806

2 People who think of retreating before a battle has been fought ought to have stayed at home.
Marshal Michel Ney, to Marshal Jomini before the Battle of Elchingen, 1805

3 Two conditions remain absolutely imperative: the infantry must be good, swift marchers, accustomed to fatigue; also their fire power must be effective.
Marshal Michel Ney, Order of the Day before the Battle of Elchingen, 1805

4 Soldiers, when I give the command to fire, fire straight at my heart. Wait for the order. It will be my last to you.
Marshal Michel Ney, last words before his execution by firing squad, 1815

5 What a soldier! The army of France is full of brave men, but Michel Ney is truly the bravest of the brave.
Napoleon I, during the retreat from Moscow, 1812. (The title 'bravest of the brave' had been bestowed by him, with equal reason, on Marshal Lannes.)

---

107 NIGHTINGALE
Florence, reformer of military nursing (1820–1910)

---

1 Do not engage in any paper wars.
Florence Nightingale, letter to Lothian Nicholson, 1853

2 You Gentlemen of England who sit at home in all the well-earned satisfaction of your successful cases, can have little idea from reading the newspapers of the Horror and Misery of operating upon these dying, exhausted men.
Florence Nightingale, letter to Dr Brown from Scutari, during the Crimean War, 1854

3 I have never seen so teachable and helpful a class as the army generally.
Florence Nightingale, letter to her sister, 1856

4 When all the medical officers have retired for the night, and silence and darkness have settled down upon those miles of prostrate sick, she may be observed alone, with a little lamp in her hand, making the solitary rounds.
John Cameron MacDonald, report in *The Times* from Scutari describing the work of Florence Nightingale, 1855

5 A Lady with a Lamp shall stand
In the great history of the land,
A noble type of good
Heroic womanhood.
Henry Wadsworth Longfellow, *Santa Filomena* (1858)

108 OWAIN GLYN DWR
('Owen Glendower'), Prince of Wales
(c. 1354–1416)

1 Owain went into hiding on St Matthew's
Day in Harvest and thereafter his hiding
place was unknown. Very many say that
he died; the seers maintain that he did
not.
Gruffudd Hiraethog, Penarth
Manuscript, 1415

2 Like a second Assyrian, the rod of God's
anger, he did deeds of unheard-of cruelty
with fire and sword.
Adam of Usk, *Chronicon Adae de Usk*
(c. 1420)

3   At my nativity
The front of heaven was full of fiery shapes,
Of burning cressets, and at my birth
The frame and huge foundation of the earth
Shak'd like a coward. (Glendower)
William Shakespeare, *Henry IV Part I,* III, i

4 This desolation arose from Owen Glyndwr's
policie, to bring all things to waste, that the
English should find no strength, nor resting
place.
John Wynn, *History of the Gwydir Family*
(1770)

109 PARMA
Alessandro Farnese, Prince of Parma,
Spanish soldier (1545–92)

1 God will grow weary of working miracles
for us.
Alessandro Farnese, Prince of Parma, letter
to King Philip II of Spain after the Siege of
Grave, 1586, quoted in *Papeles de la
secretario de estado* (1586)

2 I am only a man and cannot work miracles.
Alessandro Farnese, Prince of Parma, letter
to King Philip II of Spain, after the failure of
his planned invasion of England, 1588,
quoted in *Papeles de la secretario de estado*
(1588)

3 I know I have the body of a weak and feeble
woman, but I have the heart and stomach of
a king, and of a king of England too, and
think foul scorn that Parma or Spain or any
prince of Europe should dare to invade the
borders of my realm.
Queen Elizabeth I of England, address to
the English Army at Tilbury before Parma's
planed invasion of England, 1588

110 PATTON
George Smith, American general
(1885–1945)

1 Wars may be fought with weapons, but they
are won by men. It is the spirit of the men
who follow and of the man who leads that
gains the victory.
General George S. Patton, in the *Cavalry
Journal,* 1933

2 War will be won by Blood and Guts alone.
General George S. Patton, address to fellow
officers, Fort Benning, Georgia, 1940

3 We shall attack and attack until we are
exhausted, and then we shall attack again.
General George S. Patton, address to his
troops before the invasion of North Africa,
1942

4 The most vital quality a soldier can possess
is self-confidence, utter, complete and
bumptious.
General George S. Patton, letter to his son,
1944

5 Practically everyone but myself is a
pusillanimous son of a bitch.
General George S. Patton, letter to Colonel
Codman, 1945

6 In war nothing is impossible, provided you
use audacity.
General George S. Patton, *War As I Knew It*
(1947)

7 Never tell people how to do things. Tell
them what to do and they will surprise you
with their ingenuity.
General George S. Patton, *War As I Knew It*
(1947)

8 The greatest weapon against the so-called
'battle fatigue' is ridicule.
General George S. Patton, *War As I Knew It*
(1947)

9 Patton should have lived during the

Napoleonic wars — he would have made a splendid Marshal under Napoleon.
Field Marshal Earl Alexander, *Memoirs* (1962)

111 PERICLES
Athenian statesman and war leader
(*c.* 494–429 BC)

1 Athenians, my views are the same as ever: I am against making any concessions to the Peloponnesian, even though I am aware that the enthusiastic state of mind in which people are persuaded to enter upon a war is not retained when it comes to action.
Pericles, speech in reply to the Spartans' ultimatum, 432 BC, quoted in Thucydides, *The Peloponnesian Wars*, I, 140, trs. Warner

2 In war opportunity waits for no man.
Pericles, speech in reply to the Spartans' ultimatum, 432 BC, quoted in Thucydides, *The Peloponnesian Wars*, I, 142, trs. Warner

3 One's sense of honour is the only thing that does not grow old.
Pericles, funeral oration to the Athenian dead, 431 BC, quoted in Thucydides, *The Peloponnesian Wars*, II, 44, trs. Warner

4 The man who can most truly be accounted brave is he who best knows the meaning of what is sweet in life and of what is terrible, and then goes out undeterred to meet what is to come.
Pericles, funeral oration to the Athenian dead, 431 BC, quoted in Thucydides, *The Peloponnesian Wars*, II, 39, trs. Warner

5 Trees, though they are cut and lopped, grow up again quickly, but if men are destroyed, it is not easy to get them up again.
Pericles, quoted in Plutarch, *Lives*, 33

6 No Athenian ever put on mourning because of me.
Pericles, on his deathbed, 429 BC, quoted in Plutarch, *Lives*, 38

7 For then in wrath the Olympian Pericles Thundered and lightened and confounded Hellas.
Aristophanes, *Frogs*, trs. Rogers

8 Such a bitter smoke ascended while the flames of war he blew

That from every eye in Hellas everywhere the tears it drew.
Aristophanes, *Peace*, trs. Rogers

112 PERSHING
John Joseph, American general
(1860–1948)

1 He's a fighter — a fighter — a fighter.
General John J. Pershing, favourite words of praise while commanding the American Expeditionary Force in France, 1918

2 A competent leader can get efficient service from poor troops, while on the contrary an incapable leader can demoralise the best of troops.
General John Pershing, *Experiences in the World War*, vol. II (1931)

3 When the last bugle is sounded, I want to stand up with my soldiers.
General John Pershing, quoted in Hale and Turner, *The Yanks Are Coming* (1983)

4 I suppose our campaigns are ended, but what an enormous difference a few more days would have made!
General John Pershing, on the Armistice, 1918, quoted in Griscom, *Diplomatically Speaking* (1940)

5 Oh to be in Paris now that Pershing's there!
Anon., music-hall song celebrating his arrival in France as head of the United States Expeditionary Force, 1917

113 PÉTAIN
Henri Philippe Omer, Marshal of France
(1856–1951)

1 The present war has taken the form of a war of attrition. There is no decisive battle as there used to be. Success will belong finally to the side which possesses the last man.
Marshal Philippe Pétain, note on operations in the Arras region, 1915, quoted in Bourget, *Un certain Philippe Pétain* (1966)

2 The method of wearing out the enemy while suffering the minimum casualties oneself consists in multiplying limited attacks, mounted with great artillery support, so as

to strike the vault of the German edifice
without pause until it collapses.
Marshal Philippe Pétain, on becoming
French Commander-in-Chief, 1917, quoted
in Serrigny, *Trente Ans avec Pétain* (1959)

3   The Germans will beat the English *en rase
campagne*; after which they will beat us as
well.
Marshall Philippe Pétain, statement at
Allied conference in Doullens, 1918, quoted
in Clemenceau, *Grandeurs et Misères d'une
Victoire* (1930)

4   The true chief is one who knows how to ally
firmness with wisdom, professional
knowledge with resolution in action, the art
of the organiser with that of the executor. It
is thus that he wins confidence.
Marshal Philippe Pétain, *Le Devoir des
Elites* (1936)

5   It is easy to prove to the German people that
they have not been beaten. I only hope that
that does not lead us into a second World
War even more terrible than the first!
Marshal Philippe Pétain, letter to Mme
Pardée, 1938

6   I make to France the gift of my person.
Marshal Philippe Pétain, statement to the
people of France following the surrender to
Germany, 1940

7   I found him businesslike, knowledgeable
and brief of speech. The latter is, I find, a
rare quality in a Frenchman!
Field Marshal Earl Haig, on meeting Pétain,
1916

8   Pétain did not appear to me only as a
soldier; his greatness does not only derive
from his skill at directing a battle, but
emanates from his entire personality.
Jean de Pierrefeu, *GQG — Secteur I* (1920)

9   A soldier before all, and one with strong will
and decided opinions.
Charles à Court Repington, *The First World
War*, vol. I (1920)

114   PETER I
(‘Peter the Great’) Tsar of Russia
(1672–1725)

1   It is your military genius that has inspired
my sword, and the noble emulation of your

exploits has aroused in my heart the first
thoughts I ever had of enlarging my Empire.
Tsar Peter I, letter to King William III of
Britain, 1697

2   I have not spared and I do not spare my life
for my fatherland and people.
Tsar Peter I, letter to his son Alexis, 1715

3   Two things are necessary in government,
namely order and defence.
Tsar Peter I, letter to his son Alexis, 1715

4   Ah, lord of doom
And potentate, 'twas thus, appearing
Above the void, and in thy hold
A curb of iron, thou sat'st of old
O'er Russia, on her haunches rearing.
Alexander Pushkin, *The Bronze Horseman*
(1833), trs. Elton

5   The impact of Peter the Great upon
Muscovy was like that of a peasant hitting
his horse with his fist.
B.H. Sumner, *Survey of Russian History*
(1944)

115   POMPEY
Gnaius Pompeius (‘Pompey the Great’),
Roman soldier (104–45 BC)

1   For in whatever part of Italy I stamp the
earth with my foot, there will spring up
forces, both footsoldiers and horsemen.
Pompey, quoted in Plutarch, *Lives*

2   A dead man cannot bite.
Pompey, quoted in Plutarch, *Lives*

3   More worship the rising than the setting
sun.
Pompey, quoted in Plutarch, *Lives*

4   If only Pompey had died before the Civil
War broke out . . . he would have taken to
the grave unimpaired all those high qualities
he had borne throughout his life.
Velleius Paterculus, *Roman Histories*, II, 48

116   PONTIAC
Chief of the Ottawa (1720–69)

1   When you are going to war, you juggle, join

the medicine dance, and believe that I am speaking.
Chief Pontiac, speech to war council, Detroit River, 1763

2 I mean to destroy the English and leave not one upon our lands.
Chief Pontiac, quoted in Eckert, *The Conquerors* (1972)

3 He possessed a commanding energy and force of mind, and in subtlety and craft could match the best of his wily race.
Frances Parkman, *The Conspiracy of Pontiac* (1870)

117 QUINTUS FABIUS MAXIMUS
Roman consul and warleader (d. 203 BC)

1 In this I bring you war and peace: take which you choose.
Quintus Fabius Maximus, Order to the Carthaginians, demanding reparation for Saguntum, 218 BC, quoted in Livy, *Histories,* XXI, 19

2 Let us leave the Gods here: they are angry with the people of Tarentum.
Quintus Fabius Maximus, address to the Roman army after the capture of Tarentum, 209 BC

3 To be turned from one's course by one's opinions, by blame and by misrepresentation shows a man unfit to hold office.
Quintus Fabius Maximus, quoted in Plutarch, *Lives,* 5

4 By delaying he preserved the state.
Quintus Ennius, of the tactics which saved Rome, quoted in Cicero, *De Senectute,* IV

5 (Of his appointment as dictator, 233 BC) He was determined not to fight a pitched battle, and since he had time and manpower and money on his side, his plan was to exhaust his opponents' strength which was now at its peak by means of delaying tactics, and gradually wear down his small army and meagre resources.
Plutarch, *Lives.*

118 RAGLAN
Fitzroy James Henry Somerset, Baron Raglan, British field marshal (1788–1855)

1 Hallo! Don't carry away that arm till I've taken off my ring.
Lord Raglan, after losing his arm at the Battle of Waterloo, 1815

2 He is a good brave soldier, I am sure, and a polished gentleman, but he is no more fit than I am to cope with any leader of strategic skill.
William Howard Russell, letter to John Thadeus Delane, 1854

3 He was dignified, brave, courteous, gentle and honourable to the point of saintliness. Unfortunately these virtues do not make for success or survival in a tough world.
Correlli Barnett, *Britain and Her Army* (1970)

119 RICHARD I
('Richard the Lion Heart') King of England (1157–99)

1 Dear Lord, I pray Thee to suffer me not to see Thy Holy City, since I cannot deliver it from the hands of Thy enemies.
King Richard I, attr. words on catching sight of Jerusalem, 1192

2 When one has a good reserve, one does not fear one's enemies.
King Richard I, after his victory at the Battle of Freteval, 1194, quoted in *Histoire de Guillaume le Maréchal* (c. 1220)

3 Friends have I many, but their gifts are slight;
Shame to them if unransomed I, poor wight,
Two winters languish here!
King Richard I, poem written in captivity, 1194, quoted in Norgate, *Richard the Lionheart* (1924)

4 As regards his kingdom and rank he was inferior to the King of France but he outstripped him in wealth, in valour and in fame as a soldier.
Beha al-Din, 1190, quoted in *Recueil des Historiens des Croisades* (1872–1906)

5 Richard's courage, shrewdness, energy and patience made him the most remarkable ruler of his times.
Ibn al-Athir, on King Richard I's death, 1199, quoted in *Recueils des Historiens des Croisades* (1872–1906)

6 Then King Richard, fierce and alone, pressed on the Turks, laying them low; none whom his sword touched might escape; for wherever he went he made a wide path for himself, brandishing his sword on every side.
Ambrose, *Itinerarium Ricardi* (c. 1200)

7 If heroism be confined to brutal and ferocious valour, Richard Plantaganet will stand high among the heroes of the age.
Edward Gibbon, *Decline and Fall of the Roman Empire* (1776–88)

8 Richard I was rather a knight-errant than a king. His history is more that of a Crusade than a reign.
Sir James MacKintosh, *History of England* (1830–2)

9 Coeur-de-Lion was not a theatrical popinjay with greaves and steel-cap on it, but a man living upon victuals.
Thomas Carlyle, *Past and Present* (1843)

120 RICHARD III
King of England (1452–85)

1 What prevaileth a handful of men to a whole nation? As for me, I assure you this day I will triumph by glorious victory or suffer death for immortal fame.
King Richard III, address to his army before the Battle of Bosworth, 1485

2 Give me my battle axe in my hand
And set my crown on my head so high!
For by Him that made both Sun and Moon,
King of England I will die.
Humphrey Brereton, 'The Most Pleasant Song of the Lady Bessy', (c. 1486)

3 Frende and foo was muche what indifferent, where his aduantage grew, he spared no man's deathe, wore life withstode his purpose.
Sir Thomas More, *The Historie of Kynge Rycharde the Thirde* (1543)

4 Give me another horse! bind up my wounds!
Have mercy Jesu! Soft! I did but dream.
O coward conscience, how dost thou afflict me!
William Shakespeare, *Richard III*, V, iii

5 A horse! a horse! my kingdom for a horse!
William Shakespeare, *Richard III*, V, iv

121 RIDGWAY
Matthew Bunker, American general
(1895–   )

1 There is still an absolute weapon: that weapon is man.
General Matthew B. Ridgway, speech at Cleveland, Ohio, 1953

2 The soldier is the statesman's junior partner.
General Matthew B. Ridgway, *The Soldier and the Statesman* (1954)

3 Professional soldiers are sentimental men, for all the harsh realities of their calling. In their wallets they carry bits of philosophy, fragments of poetry, quotations from the Scriptures which in time of stress and danger speak to them with great meaning.
General Matthew B. Ridgway, *My Battles in War and Peace* (1956)

4 Every soldier learns in time that war is a lonely business.
General Matthew B. Ridgway, *The War in Korea* (1967)

5 One of the major mistakes in Korea was our tendency to try to base our strategy on a reading of enemy intentions, while failing to give proper weight to what we knew of enemy capabilities.
General Matthew B. Ridgway, *The War in Korea* (1967)

6 What red-blooded American could oppose so shining a concept as victory? It would be like standing up for sin against virtue.
General Matthew B. Ridgway, *The War in Korea* (1967)

7 He possessed almost all the military virtues — courage, brains, ruthlessness, decision.
Max Hastings, *The Korean War* (1987)

122 ROBERT I
('Robert the Bruce') King of Scotland
(1274–1329)

1 He, that his people and heritage might be
delivered out of the hands of our enemies,
met toil and fatigue, hunger and peril, like
another Maccabeus or Joshua, and bore
them cheerfully.
Bernard de Linton, letter to Pope John
XXII, 1320, also known as the Declaration
of Arbroath

2 For me think that richt speidfull
[advantageous] war,
To gang on fut to this fechting,
Armyt bot in-to licht armyng.
John Barbour, *The Bruce* (*c.* 1375)

3 (Address to the Scottish army before
Bannockburn)
Yhe mycht haf liftit-in-to thrildome;
Bot, for ye yarnit till haf fredome,
Yhe ar assemblit heir with me;
Tharfor is neidfull that yhe be
Worthy and wicht, but abaysyng.
John Barbour, *The Bruce* (*c.* 1375)

4 Wha for Scotland's King and Law
Freedom's sword will strongly draw,
Free-man stand, or free-man fa'
Let him follow me!
Robert Burns, 'Robert Bruce's Address to
his Army before Bannockburn'

5 His was the patriot's burning thought,
Of Freedom's battle bravely fought.
Sir Walter Scott, *The Lord of the Isles*, III
(1815)

123 ROBERTS
Frederick Sleigh, Earl Roberts of
Kandahar, British field marshal
(1832–1914)

1 Before a man can be manly the gift which
makes him so must be there, collected by
him slowly, unconsciously as his bones, his
flesh, his blood.
Field Marshal Earl Roberts, Diary, 1876

2 (Of the march from Kabul to Kandahar) I
fancy myself crossing and recrossing the
river which winds through the pass; I hear
the martial beat of drums and plaintive
music of the pipes; and I see Riflemen and
Gurkhas, Highlanders and Sikhs, guns and
horses, camels and mules, with the endless
following of the Indian army, winding
through the narrow gorges, or over the
interminable boulders which made the
passage of the Bolan so difficult and
wearisome to man and beast.
Field Marshal Earl Roberts, *Forty-One
Years in India*, II (1897)

3 To the discipline, bravery and devotion to
duty of the Army in India, in peace and war,
I felt that I owed whatever success it was my
good fortune to achieve.
Field Marshal Earl Roberts, *Forty-One
Years in India*, II (1897)

4 I feel the greatest hesitation and dislike to
expressing my opinion thus plainly, and
nothing but the gravity of the situation and
the strongest sense of duty would induce me
to do so, or to offer — as I now do — to
place my services at the disposal of the
Government.
Field Marshal Earl Roberts, letter to
Lord Lansdowne, offering to take
command of the British forces in South
Africa, 1899

5 Arm and prepare to quit yourselves like
men, for the time of your ordeal is at hand.
Field Marshal Earl Roberts, *Message to the
Nation* (1912)

6 There's a little red-faced man,
Which is Bobs,
Rides the tallest 'orse 'e can —
Our Bobs.
Rudyard Kipling, 'Bobs Bahadur' (1898)

7 And glory is the least of things
That follow this man home.
Rudyard Kipling, 'Lord Roberts'
(1914)

8 What gave him his great strength in counsel,
as in the field, was the simple modesty of his
confidence.
F.S. Oliver, *Ordeal by Battle* (1914)

9 The greatest and most lovable man I have
every known.
Maj-Gen Sir Frederick Maurice, *Life of
General Lord Rawlinson* (1928)

124 ROBERTSON
Sir William Robert, British field marshal (1869–1933)

1 'Orace, you're for 'ome.
Field Marshal Sir William Robertson,
Order to General Sir Horace
Smith-Dorrien, relieving him of the
command of the British Second Army in
France, 1915

2 I am more concerned than ever that it is
we who will have to finish this war and
therefore in every way we possibly can we
must take the lead, or at any rate refuse to
be led against our judgement.
Field Marshal Sir William Robertson,
letter to Field Marshal Earl Haig, 1916

3 It is very hard work trying to win this war.
The Boche gives me no trouble compared
with what I meet in London.
Field Marshal Sir William Robertson,
letter to Lord Northcliffe, 1916

4 It's a waste of time explaining strategy to
you. To understand my explanation you
would have had to have my experience.
Field Marshal Sir William Robertson,
remark to David Lloyd George, after
becoming Chief of the Imperial General
Staff, 1916

5 Details, so called, were thought to be
petty and beneath the notice of
big-minded men, and yet they are the very
things which nine hundred and ninety
times out of a thousand make just the
difference in war between success and
failure.
Field Marshal Sir William Robertson,
From Private to Field Marshal (1921)

6 It was not altogether agreeable to be seen
drinking water at mess when others were
drinking champagne.
Field Marshal Sir William Robertson,
From Private to Field Marshal (1921)

7 Modern war, being largely a matter of
war against economic life, it has turned
more and more towards the enemy's
home country, and the old principle of
making war only against armies and
navies has been consigned to the
background.
Field Marshal Sir William Robertson,
From Private to Field Marshal (1921)

8 I would rather Bury you than see you in a
red coat.
Mrs Ann Robertson, letter to her son
William Robertson after he joined the
16th Lancers, 1877

9 He never really liked foreigners and he
found most of his civilian colleagues a
trial. If they were ignorant of military
matters, as they usually were, he had no
patience with them.
Victor Bonham-Carter, *Soldier True*
(1963)

125 RODRIGO DIAZ DE VIVAR
El Cid Campeador ('El Cid'), Castilian soldier (*c.* 1043–99)

1 All I can say is that I shall fight man to man
and leave the rest to God.
El Cid, before the Battle of Golpejerra,
1072, quoted in **Romancero de Cid**
(1612)

2 Now enough of brave words; let us settle
our differences by arms, like men.
El Cid, challenge to Count Berenguer of
Barcelona before the Battle of Pinar de
Tevar, 1090, quoted in *Historia Roderici*
(*c.* 1122)

3 You see my sword dripping with blood and
my horse sweat. It is thus that the Moors are
conquered in the field of battle.
El Cid, after the rout of the Almoravides
before Valencia, 1094, quoted in *Cronica
Particular del Cid* (1512)

4 God has promised me a great victory after
my death.
El Cid, last words before his death, 1099,
prophesying the lifting of the Siege of
Valencia, quoted in *Cronica del Cid Ruy
Diaz* (1498)

5 This man, the scourge of his time, was yet
by his love of glory, his firm strength of
character, and his heroic valour, the miracle
of God's miracles.
Ibn Bassam, 'Treasury of the Excellencies of
the Spaniards' *c.* 1110, quoted in Dozy,
*Recherches sur l'histoire politique de
L'Espagne pendant le Moyen Age* (1849)

126 ROMMEL
Erwin, German field marshal
(1891–1944)

1 War makes extremely heavy demands on the
soldier's strength and nerves. For this
reason, make heavy demands on your men
in peacetime.
Field Marshal Erwin Rommel, *Infantry
Attacks* (1937)

2 To the best of my belief I have done all I
could for victory.
Field Marshal Erwin Rommel, letter to his
wife before the Battle of El Alamein, 1942

3 What will history say in passing its verdict
on me? If I am successful here, then
everybody else will claim all the glory — just
as they are already claiming the credit for
the defences and the beach obstacles that I
have erected. But if I fail here, then
everybody will be after my blood.
Field Marshal Erwin Rommel, Diary, 1944

4 The commander must be the prime mover of
the battle and the troops must always have
to reckon with his appearance in personal
control.
Field Marshal Erwin Rommel, quoted in
Liddell Hart, *The Rommel Papers* (1953)

5 The commander must establish personal
and comradely contact with his men but
without giving away an inch of his
authority.
Field Marshal Erwin Rommel, quoted in
Liddell Hart, *The Rommel Papers* (1953)

6 The ordinary soldier has a surprisingly good
nose for what is true and what is false.
Field Marshal Erwin Rommel, quoted in
Liddell Hart, *The Rommel Papers* (1953)

7 Prejudice against innovation is a typical
characteristic of an Officer Corps which has
grown up in a well-tried and proven system.
Field Marshal Erwin Rommel, quoted in
Liddell Hart, *The Rommel Papers* (1953)

8 We have a very daring and skilful opponent
against us, and, may I say across the havoc
of war, a great general.
Winston S. Churchill, speech to House of
Commons, 1942

9 Accept my sincerest sympathy for the heavy
loss you have suffered with the death of
your husband. The name of Field Marshal

Rommel will be for ever linked with the
heroic battles in North Africa.
Adolf Hitler, letter to Rommel's wife after
he had been forced to commit suicide, 1944

10 Utterly fearless, full of drive and initiative,
he was always up in front where the battle
was fiercest. If his opponent made a
mistake, Rommel was on to it like a flash.
Lt-Gen Sir Brian Horrocks, *A Full Life*
(1960)

11 He was a first-class commander; you
couldn't go to sleep when he was about.
Field Marshal Sir Claude Auchinleck,
interview with David Dimbleby, 1976

12 My father had three ambitions for me: he
wanted me to become a fine sportsman, a
great hero and a good mathematician. He
failed on all three counts.
Manfred Rommel, quoted in Irving, *The
Trail of the Fox* (1977)

127 RUPERT
Count Palatine of the Rhine, German
soldier (1619–82)

1 I think there is none that take me for a
Coward; for sure I feare not the face of any
man alive.
Prince Rupert, *Prince Rupert His
Declaration* (1642)

2 One comfort will be left: we shall all fall
together. When this is, remember I have
done my duty.
Prince Rupert, letter to the Duke of
Richmond during the siege of Bristol, 1645

3 A man who hath his hand very deep in the
blood of innocent men.
Oliver Cromwell, letter to David Leslie,
1650

4 The Prince was rough and passionate, and
loved not debate, like what was proposed,
as he liked the person who proposed it.
Earl of Clarendon, *History of the Rebellion*
(1704)

5 His very name struck terror in Puritan
hearts.
Mark Bence-Jones, *The Cavaliers* (1976)

128 SALADIN
Sultan of Syria and Egypt (1138–93)

1 I do not want to lay down my arms until there is no longer a single infidel on earth, unless between now and that time death prevents me.
Saladin, quoted by Beha ed-Din, 1171, in *Recueil des Historiens des Croisades* (1872–1906)

2 My saddle is my council chamber.
Saladin, quoted by Ibn al-Athir, *c.* 1180, in *Recueil des Historiens des Croisades* (1872–1906)

3 Kings do not kill each other. It is not the custom.
Saladin, after the defeat of King Guy's Crusader army, 1187, quoted by Imad al-Din in *Recueil des Historiens des Croisades* (1872–1906)

4 I should like to know what advantage it would be to you to lose the support of a man like me.
Saladin, letter to Imad al-Din Zenghi of Sinjar, 1190

5 I take God to witness I would rather lose all my children than cast a single stone from the walls, but God wills it; it is necessary for the Moslem cause, therefore I am obliged to carry it through.
Saladin, order to his army before the destruction of Ascalon, 1191

6 I would rather be gifted with wealth, so long as it is accompanied by wisdom and moderation, than with boldness and immoderation.
Saladin, at the end of the Third Crusade, 1192, quoted by Beha al-din in *Recueil des Historiens des Croisades* (1872–1906)

7 Wise in counsel, valiant in war and generous beyond measure.
William of Tyre, *Historia Rerum in Partibus Transmarinis Gestarum* (*c.* 1180)

129 SAXE
Hermann Maurice, Comte de Saxe, German soldier in French service (1696–1750)

1 I cannot betray a people to whom I have given my promise. I cannot dishonour myself in the eyes of a nation which has placed its trust in me.
Marshal Maurice de Saxe, letter to his half-brother, King Augustus III of Poland, pledging his support, 1726

2 No need to write long messages to a good soldier. Fight on. I am on my way.
Marshal Maurice de Saxe, field message to the Marquis de Courtivon before the Siege of Egra, 1742

3 The circumstances of war are sensed rather than explained; and, if war depends on inspiration, it seems foolish to stick pins into the soothsayer.
Marshal Maurice de Saxe, memorandum before the Battle of Laufeldt, 1747

4 Decline the attack altogether unless you can make it with advantage.
Marshal Maurice de Saxe, *Mes Rêveries* (1757)

5 The human heart is the starting point in all matters pertaining to war.
Marshal Maurice de Saxe, *Mes Rêveries* (1757)

6 I am persuaded that unless troops are properly supported in action, they will be defeated.
Marshal Maurice de Saxe, *Mes Rêveries* (1757)

7 It is certain that the French army reaped no profit from its experience of Maréchal de Saxe, and the high theatricalities, ornamental blackguardisms, and ridicule of life and death. In the long run a graver face would have been of better augury.
Thomas Carlyle, *History of Frederick the Great,* VI (1858–65)

8 That connoisseur of the art of war.
Captain Sir Basil Liddell Hart, *Thoughts on War* (1944)

130 SCHARNHORST
Gerhard Johann David von, Prussian general (1755–1813)

1 As to the Prussian Army, it is animated by the best spirit. With courage and ability nothing is wanting.
Gerhard von Scharnhorst, letter to his son, 1805

2 What we ought to do I known right well.
What we *shall* do God only knows.
Gerhard von Scharnhorst, before the Battle
of Auerstadt, 1806

3 Are noblemen's children to have the
privilege of being appointed officers in their
crass ignorance and feeble childhood while
educated and energetic men are set below
them without hope of promotion? So much
the better, no doubt, for the noble families,
but ill for the army.
Gerhard von Scharnhorst, *Comparison
between the former and the present conduct
of business in the upper part of the Military
Department* (1809)

4 Good marksmanship is always the most
important thing for the infantry.
Gerhard von Scharnhorst, memorandum,
1812

5 Better to lose another battle than to lose
Scharnhorst.
Marshal Blücher, on the death of
Scharnhorst, 1813

---

131 SCHLIEFFEN
**Alfred, Graf von Schlieffen, German field
marshal (1833–1913)**

---

1 The enemy's front is not the objection. The
essential thing is to crush the enemy's flanks
. . . and complete the extermination by
attack upon his rear.
Field Marshal Count von Schlieffen,
Cannae (1913)

2 It must come to a fight. Only make the right
wing strong.
Field Marshal Count von Schlieffen, dying
words, 1913

3 (Of the Schlieffen Plan) In our unfavourable
military-political situation in the centre of
Europe, surrounded by enemies, we had to
reckon with foes greatly superior in numbers
and prepare ourselves accordingly, if we did
not wish to allow ourselves to be crushed.
General Erich von Ludendorff, *My War
Memoirs* (1920)

4 In 1914 his dead hand automatically pulled
the trigger.
A.J.P. Taylor, *The First World War* (1963)

---

132 SCIPIO AFRICANUS
**Publius Cornelius Scipio, Roman general
(237–183 BC)**

---

1 Keep this fact before your eyes: that if you
overcome the enemy not only will you be
the complete masters of Africa, but you will
win for yourselves and for Rome the
unchallenged leadership and sovereignty of
the rest of the world.
Publius Cornelius Scipio Africanus, speech
to his army before the Battle of Zama,
202 BC, quoted in Polybius, *Histories,* XV,
10, trs. Scott-Kilvert

2 Ungrateful country, you will not possess
even my bones.
Publius Cornelius Scipio Africanus, words
to be left on his tomb in Campania, after
being tried for bribery in Rome, quoted in
Valerius Maximus, *De Dictis Factisque,* V, 3

3 He was careful to refrain from exposing
himself to danger when his country's entire
hopes rested upon his safety. Such conduct
is not the mark of a general who trusts to
luck, but of one who possesses intelligence.
Polybius, *Histories,* X, 6, trs. Scott-Kilvert

4 Heaven itself seems to have formed this
particular hero to mark out to the rulers of
this world the art of governing with justice.
Abbé Seran de la Tour, *Scipio* (1739)

5 Scipio is the embodiment of grand strategy,
as his campaigns are the supreme example
in history of its meaning.
Captain Sir Basil Liddell Hart, *A Greater
than Napoleon* (1930)

---

133 SCIPIO AFRICANUS THE YOUNGER
**Publius Cornelius Aemilianus Scipio,
Roman general (185–129 BC)**

---

1 I do not like the enthusiastic man.
Cornelius Scipio Africanus Aemilianus,
quoted in Cicero, *De Oratore,* II, 67

2 It is a fine thing, but I have an
unaccountable fear and dread lest some day
someone else should give this same order to
my own city.
Cornelius Scipio Africanus Aemilianus, on

giving the order to set fire to Carthage, 146 BC, quoted in Polybius, *Histories*, XXXVIII, 21

3 When I am at leisure I do most work.
Cornelius Scipio Africanus Aemilianus, quoted in Plutarch, *Scipionis Apophthegmata*, I

4 Rome cannot fall while Scipio stands, nor Scipio live when Rome has fallen.
Cornelius Scipio Africanus Aemilianus, speech to the Gracchan mob before his death, 129 BC, quoted in Plutarch, *Scipionis Apophthegmata*, XXIII

5 Never less idle than when unoccupied, never less alone than when without company.
Cicero, *De Officiis*, III, i

134 SHERMAN
William Tecumseh, American (Union) general (1820–91)

1 If we must be enemies, let us be men, and fight it out as we propose to do, and not deal in hypocritical appeals to God and Humanity.
General William T. Sherman, letter to General Hood, Atlanta, Georgia, 1864

2 Strengthen your position; fight anything that comes.
General William T. Sherman, Order to Maj-Gen Macpherson before Atlanta, 1864

3 You cannot qualify wars in harsher terms than I will. War is cruelty, and you can't refine it.
General William T. Sherman, letter to James M. Calhoun, 1864

4 There is a soul to an army as well as to the individual man, and no general can accomplish the full work of his army unless he commands the soul of his men as well as their bodies and legs.
General William T. Sherman, *Memoirs* (1875)

5 To be at the head of a strong section of troops, in the execution of some task that requires brain, is the highest pleasure of war.
General William T. Sherman, *Memoirs* (1875)

6 Every attempt to make war easy and safe will result in humiliation and disaster.
General William T. Sherman, quoted in Keim, *Sherman: A Memorial in Art, Oratory and Literature* (1904)

7 When we come to fight Indians, I will take my code from soldiers and not from citizens.
General William T. Sherman, quoted in Sell and Weybright, *Buffalo Bill and the Wild West* (1959)

8 There's many a boy here today who looks on war as all glory; but, boys, it is all hell.
General William T. Sherman, speech at Columbus, Ohio, 1880

9 You have accomplished the most gigantic undertaking given to any general in this war, and with a skill and ability that will be acknowledged in history as unsurpassed if not unequalled.
General Ulysses S. Grant, letter to Sherman after the capture of Atlanta, 1864

10 We drink to twenty years ago.
When Sherman led our banner;
His mistresses were fortresses,
His Christmas gift — Savannah!
George B. Corkhill, Toast to Sherman, Washington, 1883

135 SITTING BULL
Sioux chief (1834–90)

1 A warrior
I have been.
Now
it is all over.
A hard time
I have.
Chief Sitting Bull, 'Song of Sitting Bull'

2 Go back home where you came from. This country is mine and I intend to stay here, and to raise this country full of grown people.
Chief Sitting Bull, quoted in *Report of the Commission of Indian Affairs to the Secretary of State* (1876)

3 First kill me before you take possession of my fatherland.
Chief Sitting Bull, statement to war council, Powder River, 1877

4 I wish it to be remembered that I was the

last man of my tribe to surrender my rifle, and this day I have given it to you.
Chief Sitting Bull, statement to Major Brotherton, Fort Buford, Canada, 1881

5   What treaty that the white man ever made with us have they kept? Not one.
Chief Sitting Bull, quoted in Johnson, *Life of Sitting Bull and History of the Indian War* (1891)

6   God made me an Indian, but not a reservation Indian.
Chief Sitting Bull, quoted in Turner, *The North West Mounted Police 1873–1893*, vol. I (1950)

136 SLIM
William Joseph Viscount Slim, British field marshal (1891–1970)

1   The dominant feeling on the battlefield is loneliness.
Field Marshal Viscount Slim, address to 10th Indian Division, 1941

2   In the British Army there are no good battalions, and no bad battalions, only good and bad officers.
Field Marshal Viscount Slim, address to Sandhurst officer cadets, 1949

3   Defeat is bitter. Bitter to the common soldier, but trebly bitter to his general.
Field Marshal Viscount Slim, *Defeat into Victory* (1956)

4   The war in Burma was a *soldiers'* war.
Field Marshal Viscount Slim, *Defeat into Victory* (1956)

5   (Of the war in Burma) Victory came, not from the work of any one man, but from the sum of many men's efforts.
Field Marshal Viscount Slim, *Defeat into Victory* (1956)

6   One of the most valuable qualities of a commander is a flair for putting himself in the right place at the vital time.
Field Marshal Viscount Slim, *Unofficial History* (1959)

7   There is a freemasonry among fighting soldiers that helps them to understand one another even if they are enemies.
Field Marshal Viscount Slim, *Unofficial History* (1959)

8   While the battles the British fight may differ in the widest possible ways, they have invariably two common characteristics — they are always fought uphill and always at the junction of two or more map sheets.
Field Marshal Viscount Slim, *Unofficial History* (1959)

9   The famous Fourteenth Army, under the masterly command of General Slim, fought valiantly, overcame all obstacles, and achieved the seemingly impossible.
Winston S. Churchill, *The Second World War*, vol. VI (1954)

137 STALIN
Josef, Soviet dictator (1879–1953)

1   Print is the sharpest and strongest weapon of our party.
Josef Stalin, speech in Moscow, 1923

2   The Pope! How many divisions has he got?
Josef Stalin, in conversation with Pierre Laval, Moscow, 1935

3   A frontier in Northern France could not only divert Hitler's forces from the East, but would at the same time make it impossible for Hitler to invade Great Britain.
Josef Stalin, letter to Winston S. Churchill requesting the creation of a second front, 1941

4   History shows that there are no invincible armies.
Josef Stalin, address to the Russian people, 1941

5   Modern warfare will be a war of engines. Engines on land, engines in the air, engines on water and under water. Under these conditions, the winning side will be the one with the greater number and the more powerful engines.
Josef Stalin, address to Politburo, Kremlin, 1941

6   Our war for the freedom of our Motherland will merge with the struggle of the peoples of Europe and America for their independence, for their democratic liberties. It will be a united front of the peoples who stand for freedom and against

enslavement and threats of enslavement by Hitler's Fascist armies.
Josef Stalin, radio broadcast to the Russian people, 1941

7 The joint effort of the Soviet, US and British Armed Forces against the German invaders, which had culminated in the latter's complete rout and defeat, will go down in history as a model military alliance between our peoples.
Josef Stalin, telegram to President Truman and Winston S. Churchill, following the German unconditional surrender, 1945

8 We have already lost four million soldiers on the field of battle, and the war is not yet won — and they are human beings you know.
Josef Stalin, to Winston S. Churchill, Yalta, 1945, quoted in Alexander, *Memoirs* (1962)

9 Like Peter the Great, Stalin fought barbarism with barbarism, but he was a great man.
Nikita Khrushchev, *Khrushchev Remembers* (1971)

10 Stalin blindly believed that Hitler would not break his pact with the USSR.
Roy Medvedev, *Let History Judge* (1971)

---

138 STILWELL
Joseph Warren ('Vinegar Joe'), American general (1883–1946)

---

1 The Limeys want us in with both feet.
General Joseph Stilwell, *Papers* (1948)

2 (Of the Burma Campaign, 1943) The Limey layout is simply stupendous, you trip over Lieutenant-Generals on every floor, most of them doing Captain's work, or none at all.
General Joseph Stilwell, *Papers* (1948)

3 I claim we got a hell of a beating. We got run out of Burma, and it is as humiliating as hell. I think we should find out what caused it and go back and retake it.
General Joseph Stilwell, quoted in Fadiman, *The American Treasury* (1955)

4 He could be as obstinate as a whole team of mules; but when he said he would do a thing he did it.
Field Marshal Viscount Slim, *Defeat into Victory* (1956)

5 We thought of him simply as good old 'Vinegar Joe'.
John Masters, *The Road Past Mandalay* (1961)

---

139 TAMERLANE
(Timur) Mongol emperor (1336–1405)

---

1 I hold the Fates bound fast in iron chains. and with my hand turn Fortune's wheel about.
Christopher Marlowe, *Tamburlaine the Great* Part 1, I, i

2 Is it not passing brave to be a king, And ride in triumph through Persepolis?
Christopher Marlowe, *Tamburlaine the Great* Part 1, II, v

3 Come, let us march against the powers of heaven, And set black streamers in the firmament, To signify the slaughter of the gods.
Christopher Marlowe, *Tamburlaine the Great* Part 2, V, iii

4 For Tamburlaine the scourge of God, must die.
Christopher Marlowe, *Tamburlaine the Great* Part 2, V, iii

5 I dare say, when Tamerlane descended from his throne built of seventy thousand skulls, and marched his ferocious battalions to further slaughter, I dare say he said, 'I want more room.'
Senator Thomas Corwen, speech on the war with Mexico, Ohio, 1847

---

140 TITO
Josip Broz, Yugoslav guerrilla leader and head of state (1892–1980)

---

1 Never before, perhaps, has a small nation paid so dearly for persuading the world that the blood shed in Yugoslavia was its own blood, not the blood of the shameful traitors whose leaders are now enjoying the hospitality of the Allied countries.
Josip Broz Tito, address at the Second Session, Anti-Fascist Council of People's Liberation, 1943

2 The suffering and sacrifices and the contribution of the Yugoslav peoples to the common Allied cause has been so great and incontrovertible that we firmly believe, on the basis of this, that our just demands will not be disputed at the peace conference.
Josip Broz Tito, address at the Third Session, Anti-Fascist Council of People's Liberation, 1945

3 Although our people have, in the recent past, suffered severely at the hands of German militarism and Fascism, we harbour no feelings of hatred towards the German people.
Josip Broz Tito, address to the United Nations General Assembly, 1960

4 The struggle for the rights of peoples, for the affirmation and free comprehensive development of nationalities, is a component part of the struggle of the working class movement for its progressive revolutionary goals.
Josip Broz Tito, speech on the 25th Anniversary of The New Yugoslavia, 1968

141 TORSTENSON
Lennart, Swedish soldier (1603–51)

1 Enemies of Sweden, tremble! As long as this man can wield a sword, as long as this blood runs in my veins, I devote to you my revenge, to death, to destruction!
Field Marshal Count Lennart Torstenson, on the death of King Gustavus Adolphus, 1634

2 For I know him to be the very man who is especially qualified to lead the whole army.
King Gustavus Adolphus, letter to Count Oxienstierna, 1632

3 In battle a lion; everywhere else a lamb.
Johannes Loccenius, *History of Sweden* (1676)

4 Every city fancies Torstenson before its walls; every one is ignorant where this thunder-threatening, lightning-bearing cloud will burst.
King Gustavus III, *Eulogy of Torstenson* (1787)

142 TROTSKY
Leon (Lev Davidovich), Russian revolutionary (1879–1940)

1 A rising of the masses of the people needs no justification. What has happened is an insurrection, and not a conspiracy.
Leon Trotsky, speech to the Congress of Soviets, St Petersburg, 1917

2 A deserter from labour is as contemptible and despicable as a deserter from the battlefield. Severe punishment to both!
Leon Trotsky, Order of the Day to the Labour Armies, 1920

3 The dictatorship of the Communist Party is maintained by recourse to every form of violence.
Leon Trotsky, *Terrorism and Communism* (1924)

4 There is no 'science' of war, and there never will be any. There are many sciences war is concerned with; war is practical art and skill.
Leon Trotsky, *How the Revolution Developed its Military Power* (1924)

5 Anyone who expects to meet a lunatic brandishing a hatchet and instead finds a man hiding a revolver in his trouser pocket is bound to feel relieved. But that doesn't prevent a revolver from being more dangerous than a hatchet.
Leon Trotsky, describing Hitler in the *Bulletin of the Opposition*, 1933

6 The vengeance of history is more terrible than the vengeance of the most powerful General Secretary.
Leon Trotsky, *Stalin* (1946)

143 TURENNE
Henri de la Tour d'Auvergne, Vicomte de Turenne, Marshal of France (1611–75)

1 Make few sieges and fight plenty of battles; when you are master of the countryside the villages will give us the towns.
Vicomte de Turenne, advice to the Prince of Condé, 1643

2 When a man has committed no mistakes in

war, he can only have been engaged in it but
for a short time.
Vicomte de Turenne, after the Battle of
Marienthal, 1645

3 The enemy came. He was beaten. I am tired.
Good night.
Vicomte de Turenne, after the Battle of
Tünen, 1658

4 If I had had a man like Turenne to second
me in my campaigns I should have been
master of the world.
Napoleon I, 1817, quoted in Gourgand,
Saint-Helène Journal inédit, II (1815–18)

---

144 VILLARS
**Claude Louis Hector, Duc de Villars,
French soldier (1653–1734)**

---

1 I am going to fight your Majesty's enemies
and I leave you in the midst of my own.
Marshal de Villars, address to King Louis
XIV before setting out for the Rhine, 1704

2 Soldiers, the only hope of the King for an
honourable peace rests with your bayonets.
We shall give no quarter and ask for none.
Marshal de Villars, address to the French
Army before the Battle of Malplaquet, 1709

3 Since the army has not been able to see
Villars die a brave man's death, it is right
that it should see him die like a Christian.
Marshal de Villars, on receiving the
sacrament after being wounded at the Battle
of Malplaquet, 1709

---

145 WALLACE
**Sir William, Scottish guerrilla leader
(c. 1270–1305)**

---

1 I haf brocht you to the ring, hop gif ye kun.
Sir William Wallace, Order to the Scottish
infantry before the Battle of Falkirk, 1298

2 A man void of pity, a robber given to
sacrilege, arson and homicide, more
hardened in cruelty than Herod, more
raging in madness than Nero, he was

condemned to a most cruel but justly
deserved death.
Matthew of Westminster, *Flowers of
History* (1305)

3 In tym of pes, mek as a maid was he;
Quhar wer approchyt the rycht Ector was
he.
Blind Harry, *The Wallace* (c. 1460)

4 The story of Wallace poured a Scottish
prejudice into my veins, which will boil
along there till the flood-gates of life shut in
eternal rest.
Robert Burns, letter to Dr Moore, 1787

5 O Wallace, peerless lover of thy land
We need thee still, thy moulding brain and
hand!
Francis Lauderdale Adams,
'William Wallace' *Songs of the Army of the
Night* (1890)

---

146 WALLENSTEIN
**Albrecht Eusebius Wenzel von, Duke of
Friedland and Mecklenburg, Czech
soldier in Imperial service (1583–1634)**

---

1 Herr Albrecht von Wallenstein will await
upon Archduke Ferdinand in the camp with
180 cuirassiers and 80 musketeers
maintained at his own cost.
Albrecht von Wallenstein, message from
Prague offering his services to the Emperor
Ferdinand II, 1617

2 Better a ruined land than a lost land.
Albrecht von Wallenstein, letter to Gerhard
von Taxis, 1625

3 Had I thought so much of my soul's
salvation as of the Emperor's service I
should deservedly never come to Purgatory,
much less to Hell!
Albrecht von Wallenstein, letter to Count
von Harrach before being given command
of the Imperial Army, 1626

4 When the lands are laid in ruins we shall
have to make peace.
Albrecht von Wallenstein, after his
dismissal from the army of the Emperor
Ferdinand II, 1630

5 There may also be seen great thirst for
honour and a striving after temporal titles

and power, whereby he will make for himself many great and injurious enemies both secret and confessed.
Johann Kepler, Wallenstein's horoscope, 1607

6 Such a hero must needs have soldiers, and where soldiers are there is war, and war is there must the innocent suffer as well as the guilty.
Hans von Grimmelshausen, *Simplicissimus* (1638)

147 WASHINGTON
George, American soldier and President of the United States (1732–99)

1 As to pay, Sir, I beg leave to assure the Congress that as no pecuniary consideration could have tempted me to accept this arduous employment at the expense of my domestic ease and happiness, I do not wish to make any profit from it.
General George Washington, on being appointed Commander-in-Chief, 1775

2 I beg that you will be particularly careful in seeing strict order observed among the soldiers, as that is the life of military discipline.
General George Washington, Order to Adam Stephen, 1775

3 When we assumed the Soldier, we did not lay aside the Citizen.
General George Washington, speech to the New York Legislature, 1775

4 The Army and the Country have a mutual Dependence upon each other and it is of the last Importance that their several Duties should be so regulated and enforced as to produce not only the greatest Harmony and good Understanding but the truest Happiness and Comfort to each.
General George Washington, letter to Thomas Wharton Jr., 1778

5 I most firmly believe that the independence of the United States never will be established till there is an Army in foot for war.
General George Washington, letter to John Mathews, 1780

6 It is our true policy to steer clear of

permanent alliance with any portion of the foreign world.
General George Washington, Farewell Address, 1796

7 My first wish would be that my military family, and the whole Army, should consider themselves as a band of brothers, willing and ready to die for each other.
General George Washington, letter to Henry Knox, 1798

8 A citizen, first in war, first in peace, and first in the hearts of his countrymen.
General Henry Lee, Resolution to Congress on Washington's death, 1799

9 To add brightness to the sun or glory to the name of Washington is alike impossible. Let none attempt it.
Abraham Lincoln, speech at Springfield, Ohio, 1842

10 On the whole his character was, in its mass, perfect, in nothing bad, in few points indifferent; and it may be truly said that never did nature and fortune combine more perfectly to make a man great.
Thomas Jefferson, *Writings,* vol. XIV (1894)

148 WAVELL
Archibald Percival, Earl Wavell of Keren, British field marshal (1883–1950)

1 A bold general may be lucky, but no general can be lucky unless he is bold.
Field Marshal Earl Wavell, *Generals and Generalship* (1941)

2 The relationship between a general and his troops is very much like that between the rider and his horse. The horse must be controlled and disciplined, and yet encouraged: he should 'be cared for in the stables as if he was worth £500 and ridden in the field as if he were not worth half-a-crown'.
Field Marshal Earl Wavell, *Generals and Generalship* (1941)

3 A knowledge of humanity, on whose peculiarities, and not those of machine, the

whole practice of warfare is ultimately based.
Field Marshal Earl Wavell, *Generals and Generalship* (1941)

4 Above them towers the homely but indomitable figure of the British Soldier; the finest all-round fighting man the world has ever seen; who has won so many battles that he never doubts of victory, who has suffered so many defeats and disasters on the way to victory that he is never greatly depressed by defeat; whose humorous endurance of time and chance lasts always to the end.
Field Marshal Earl Wavell, *The Good Soldier* (1948)

5 (On being relieved of his command in 1941) The Prime Minister's quite right. The job wants a new eye and a new hand.
Field Marshal Earl Wavell, quoted in Churchill, *The Second World War*, vol. III (1950)

6 My trouble is I am not really interested in war.
Field Marshal Earl Wavell, quoted in the Diary of General Sir Henry Pownall, 1942

7 (Of the siege of Singapore) There must be no thought of surrender. Every unit must fight it out to the end and in close contact with the enemy.
Field Marshal Earl Wavell, signal to Lt-Gen A.E. Percival (1942)

8 When things look bad and one's difficulties appear great, the best tonic is consider those of the enemy.
Field Marshal Lord Wavell, *Soldiers and Soldiering* (1953)

9 A little unorthodoxy is a dangerous thing — but without it one seldom wins battles.
Field Marshal Earl Wavell, letter to Maj-Gen Eric Dorman-Smith, 1940, quoted in Barnett, *The Desert Generals* (1960)

10 The only one who showed a touch of genius.
Field Marshal Erwin Rommel, quoted in Liddell Hart, *The Rommel Papers* (1953)

11 You always felt braver after the Chief had been to see you.
R.J. Collins, *Lord Wavell* (1947)

12 He was essentially a soldier's soldier, and takes an assured place as one of the great commanders in military history.
I.S.O. Playfair, *The Mediterranean and Middle East*, vol. II (1956)

149 WELLINGTON
Arthur Wellesley, Duke of Wellington, British field marshal and prime minister (1769–1851)

1 I know but one receipt for good health in this country, and that is to live moderately, to drink little or no wine, to use exercise, to keep the mind employed, and, if possible, to keep in humour with the world. The last is more difficult, for as you have often observed, there is scarcely a good-tempered man in India.
Duke of Wellington, letter to his brother Henry, 1802

2 In Spain I shaved myself over-night, and usually slept five or six hours; sometimes indeed only three or four, and sometimes only two. I undressed very seldom, never in the first four years.
Duke of Wellington, quoted in Rogers, *Recollections* (1859)

3 When I reflect upon the characters and attainments of some of the General Officers of this army, and consider that these are the persons on whom I am to rely to lead columns against the French Generals, and who are to carry my instructions into execution, I tremble.
Duke of Wellington, letter to Lt-Col Henry Torrens, 1810

4 None but the worst description of men enter the regular service.
Duke of Wellington, letter to Lt-Col Henry Torrens, 1811

5 We have in the service the scum of the earth as common soldiers.
Duke of Wellington, letter to the Earl Bathurst, 1813

6 (Of the British soldier) It all depends on that article whether we do the business or not. Give me enough of it and I am sure.
Duke of Wellington, 1815, quoted in Creevey, *Creevey Papers* (1903)

7 (Of the Battle of Waterloo) Hard pounding this, gentlemen: let's see who will pound the longest.
Duke of Wellington, quoted in Scott, *Paul's Letters to His Kinsfolk* (1816)

8 The history of the battle is not unlike the

history of a ball. Some individuals may recollect all the little events of which the great result is the battle won or lost; but no individual can recollect the order in which, or the exact moment at which, they occurred, which makes all the difference as to their value or importance.
Duke of Wellington, letter to John Wilson Croker after the Battle of Waterloo, 1815

9 The next greatest misfortune to losing a battle is to gain such a victory as this.
Duke of Wellington, after the Battle of Waterloo, 1815, quoted in Rogers, *Recollections* (1859)

10 (Of the British soldier) The mere scum of the earth. It is only wonderful that we should be able to make so much of them afterwards.
Duke of Wellington, 1831, quoted in Stanhope, *Notes of Conversations with the Duke of Wellington* (1888)

11 All the business of war, and indeed all the business of life, is to endeavour to find out what you don't know by what you do; that's what I call 'guessing what was at the other side of the hill'.
Duke of Wellington, quoted in Croker, *Croker Papers*, III (1884)

12 I consider nothing in this country so valuable as the life and health of the British soldier.
Duke of Wellington, quoted in Guedalla, *The Duke* (1940)

13 The Duke of Wellington brought to the post of first minister immortal fame; a quality of success which would seem to include all others.
Benjamin Disraeli, *Sybil* (1845)

14 The Duke of Wellington has exhausted nature and exhausted glory. His career was one unclouded longest day.
Obituary in *The Times*, 1852

15 This is England's greatest son,
He that gained a thousand fights,
Nor never lost an English gun.
Lord Tennyson, 'Ode on the Death of the Duke of Wellington', 1851

16 This globe has produced three beings, whose names will only perish when the earth itself shall be dissolved into its elements — a POET, an ARTIST and a MAN. Of these BRITAIN claims two, Italy one; SHAKESPEARE the POET,

MICHELANGELO the ARTIST, WELLINGTON the MAN.
Sir William Fraser, *Words on Wellington* (1889)

17 Wellington was always at his coolest in the hottest of moments.
Winston S. Churchill, *A History of the English Speaking Peoples*, vol. III (1957)

---

150 WILLIAM I
('William the Conqueror') King of England (1028–87)

1 If you bear yourselves valiantly you will obtain victory, honour and riches. If not, you will be ruthlessly butchered, or else led ignominiously captive into the hands of pitiless enemies.
King William I, before the Battle of Hastings, 1066, quoted in William of Poitiers, *Gesta Willielmi Ducis Normannorum et Regis Anglorum* (*c*. 1071), trs. Douglas and Greenaway

2 I'm confident that even if I had only ten thousand men as good as the sixty thousand I've brought with me, through their courage and God's aid he and his army will be destroyed.
King William I, letter to Robert Fitzwormack refusing his offer of help before the Battle of Hastings, 1066, quoted in William of Poitiers, *Gesta Willielmi Ducis Normannorum et Regis Anglorum* (*c*. 1071), trs. Douglas and Greenaway

3 He was gentle to the good men who loved God and stern beyond measure to those who resisted his will.
Anglo-Saxon Chronicle (1087), ed. Whitelock, Douglas and Tucker

4 England's on the anvil! — hear the hammers ring —
Clanging from the Severn to the Tyne!
Never was a blacksmith like our Norman King —
England's being hammered, hammered, hammered into line!
Rudyard Kipling, 'The Anvil'

5 He writes his name with a five-flagged spear On skies of infantry in the rear.
Charles Causley, 'At the Statue of William the Conqueror, Falaise', *Collected Poems* (1975)

151 WILLIAM I
('William the Silent') Prince of Orange (1533–84)

1 A prince is constituted by God to be the ruler of a people, to defend them from oppression and violence, as the shepherd his sheep.
Prince William I, Act of Abjuration, 1581

2 No dangers can for me and mine be compared with the base desertion of such a noble cause — the honour of God, the peace of the provinces, the freedom of the Fatherland — and the abandonment of the sacred and honourable side which I have up till now followed.
Prince William I, statement to his fellow countrymen written shortly before being assassinated, 1584

3 My God, have mercy on my soul and on my poor people.
Prince William I, last words, on being assassinated, 1584

4 The Prince is a dangerous man, subtle, politic, professing to stand by the people, to champion their interests, even against your edicts, but seeking only the favour of the mob, giving himself out sometimes to stand as a Catholic, sometimes as a Calvinist or Lutheran.
Cardinal Nicholas Granvelle, letter to the Emperor Charles V, 1567

5 As long as he lived, he was the guiding star of a whole brave nation, and when he died the little children cried in the streets.
John Lothrop Motley, *The Rise of the Dutch Republic*, VI (1856)

152 WINGATE
Orde Charles, British general (1903–44)

1 Our purpose here is to found the Jewish Army.
Major-General Orde Wingate, address to Special Night Squads, Palestine, 1938

2 Finally, knowing the vanity of man's effort and the confusion of his purpose, let us pray that God may accept our services and direct

our endeavours so that when we shall have done all we shall see the fruits of our labours and be satisfied.
Major-General Orde Wingate, Order of the Day, Imphal, 1943

3 It is elementary in war to exploit success.
Major-General Order Wingate, letter to Chiefs of Staff, South-East Asia Command, 1944

4 (Of Operation Thursday, Burma Campaign, 1944) This is a moment to live in history. It is an enterprise in which every man who takes part may feel proud one day to say, 'I WAS THERE.'
Major-General Orde Wingate, Order of the Day, 1944

5 He is a man of genius and audacity and has rightly been discerned by all eyes as a figure quite above ordinary level.
Winston S. Churchill, letter to General Sir Hastings Ismay, 1943

6 A man of genius who might well have become also a man of destiny.
Winston S. Churchill, speech in House of Commons, 1944

7 He could ignite other men.
Field Marshal Viscount Slim, *Defeat into Victory* (1956)

8 (Of Wingate's death, 1944) I realised what a loss this was to the British Army and said a prayer for the soul of this man in whom I had found my match.
General Renya Mutaguchi, letter to Maj-Gen Derek Tulloch, 1968

9 At times the truth was simply not in him.
Brigadier Sir Bernard Fergusson, *The Trumpet in the Hall* (1970)

153 WOLFE
James, British general (1727–59)

1 All that I wish for myself is that I may at all times be ready and firm to meet the fate we cannot shun, and to die gracefully and properly when the time comes.
General James Wolfe, letter to his mother, 1758

2 The blame I take entirely upon my shoulders, and I expect to suffer for it.

Accidents cannot be helped. As much of the plan as was defective falls justly on me.
General James Wolfe, after the failure of his first attack on Quebec, 1759

3   We had a choice of difficulties.
General James Wolfe, words at Montmorenci Falls, 1759

4   (Of Gray's 'Elegy Written in a Country Churchyard') Gentlemen, I would rather have written those lines than take Quebec.
General James Wolfe, quoted in Parkman, *Montcalm and Wolfe* (1884)

5   To have a friend who had been near or about him was to be distinguished.
William Makepeace Thackeray, *The Virginians* (1859)

6   Oh! he is mad is he? Then I wish he would *bite* some other of my generals.
King George III, quoted in Thackeray, *History of William Pitt* (1827)

---

154   WOLSELEY
        Garnet Joseph, Viscount Wolseley,
        British field marshal (1833–1913)

---

1   My sword is thirsty for the blood of these accursed women slayers.
Field Marshal Viscount Wolseley, letter to his brother during the Indian Mutiny, 1857

2   Be cool; fire low, fire slow, and charge home; and the more numerous your enemy, the greater will be the loss inflicted upon him, and the greater your honour in defeating him.
Field Marshal Viscount Wolseley, *Soldier's Pocketbook for Field Service* (1869)

3   Surely John Bull will not endanger his birthright, his liberty, his property simply in order that men and women may cross between England and France without running the risk of sea-sickness.
Field Marshal Viscount Wolseley, memorandum on the Channel Tunnel proposals, 1882

4   When it can be said that I have appointed a bad man to an office, it will be time to find fault with my selections.
Field Marshal Viscount Wolseley, letter to his wife, 1882

5   I hope I may never return home a defeated man: I would sooner leave my old bones here than go home to be jeered at.
Field Marshal Viscount Wolseley, letter to his wife before the Battle of Tel-el-Kebir, 1882

6   My mind keeps thinking of how near a brilliant success I was, and how very narrowly I missed achieving it.
Field Marshal Viscount Wolseley, letter to his wife after Gordon's death at Khartoum, 1885

7   God must be very angry with England when he sends back Mr Gladstone to us as first minister.
Field Marshal Viscount Wolseley, letter to Lord Kitchener, 1886

8   It is often said that a man who writes well cannot be a good soldier; most of the great commanders, from King David, Xenophon, and Caesar to Wellington, not only wrote well, but extremely well.
Field Marshal Viscount Wolseley, *The Story of a Soldier's Life*, vol. II (1903)

9   I have but one great object in this world, and that is to maintain the greatness of our Empire.
Field Marshal Viscount Wolseley, quoted in Playne, *The Pre-War Mind in Britain* (1928)

10   He is one of those men who not only succeed but succeed quickly. Nothing can give you an idea of the jealousy, the hatred and all uncharitableness of the Horse Guards against our only soldier.
Benjamin Disraeli, letter to Lady Bradford, 1879

11   Wolseley was the only man I met in the army on whom command sat so easily and fitly that neither he nor the men he commanded had ever to think about it.
General Sir William Butler, *An Autobiography* (1911)

12   Wolseley by temperament and professional conviction violently attacked all that was inefficient and old-fashioned.
Correlli Barnett, *Britain and Her Army* (1970)

13   He had a good eye for military talent, and the officers he selected became known first as the Ashanti Ring and then as the Wolseley Ring or the Wolseley Gang.
Byron Farwell, *Eminent Victorian Soldiers* (1986)

155 XENOPHON
Athenian soldier and writer
(c. 430–352 BC)

1 Willing obedience always beats forced obedience.
Xenophon, *Cyropaedia*

2 For one born a man it is easier to rule all other animals than to rule men.
Xenophon, *Cyropaedia*, I, 1

3 The community of danger makes allies well disposed towards one another.
Xenophon, *Cyropaedia*, III, 2

4 With a victorious army even the camp-followers can march boldly forward.
Xenophon, *Cyropaedia*, V, 2

5 It is discipline that makes one feel safe, while lack of discipline has destroyed many people before now.
Xenophon, address to his army after the defeat of Cyrus at the Battle of Cunaxa, 401 BC, *Anabasis*, III, 1, trs. Warner

6 Nobody ever lost his life in battle from the bite or kick of a horse; it is the men who do whatever is done in battle.
Xenophon, *Anabasis*, III, 2, trs. Warner

7 Soon they heard the soldiers shouting, 'The sea, the sea', and passing the news along.
Xenophon, *Anabasis*, IV, 7, trs. Warner

8 There is a wide difference between right and wrong disposition of troops, just as stones, bricks, timber and tiles flung together anyhow are useless, whereas when the materials that neither rot nor decay, that is the stones and tiles, are placed at the bottom and at the top and the bricks and timber are put together in the middle, as in building, the result is something of great value, a house in fact.
Xenophon, *Memorabilia*, III, 1

156 XERXES
King of Persia (c. 519–465 BC)

1 May I be happy while alive and blessed when dead.
King Xerxes of Persia, Daeva inscription at Persepolis, c. 485 BC, quoted in Olmstead, *Persian Empire* (1948)

2 You salt and bitter stream, your master lays this punishment upon you for injuring him who never injured you.
King Xerxes of Persia, Order to his men to whip the waters of the Hellespont, 480 BC, quoted in Herodotus, *Histories*, VII, 36, trs. de Selincourt

3 How pitifully short human life is — for all these thousands of men not one will be alive in a hundred years' time.
King Xerxes of Persia, reviewing his army before the Hellespont, quoted in Herodotus, *Histories*, VII, 46, trs. de Selincourt

4 If we crush the Athenians and their neighbours in the Peloponnese, we shall so extend the empire of Persia that its boundaries will be God's own sky, so the sun will not look down upon any land beyond the boundaries of what is our own.
King Xerxes of Persia, address to the Persian leaders proposing the invasion of Greece, 480 BC, quoted in Herodotus, *Histories*, VII, 8, trs. de Selincourt

5 Fight this war with all your might — and for this reason: our enemies, if what I hear is true, are brave men, and if we defeat them, there is no other army in the world which will ever stand up to us again.
King Xerxes of Persia, address to his army before the invasion of Greece, 480 BC, quoted in Herodotus, *Histories*, VII, 53, trs. de Selincourt

6 Only by great risks can great results be achieved.
King Xerxes of Persia, before the invasion of Greece, 480 BC, quoted in Herodotus, *Histories*, VII, 49, trs. de Selincourt

7 Mark my words: it is through the ears you can touch a man to pleasure or rage — let the spirit which dwells there hear good things, and it will fill the body with delight; let it hear bad, and it will swell with fury.
King Xerxes of Persia, quoted in Herodotus, *Histories*, VII, trs. de Selincourt

157 YORK
Frederick Augustus, Duke of York,
British field marshal (1763–1827)

1 March to the sound of the guns.
Duke of York, address to his commanders during the Dunkirk campaign, 1793

2　I am confident you will regain my good
　opinion of you.
　Duke of York, address to his defeated
　troops before the Battle of Beaumont, 1794

3　Mercy to the vanquished is the brightest
　gem in a soldier's character.
　Duke of York, Order of the Day during the
　campaign in Flanders, 1794

4　I am not afraid of dying, I trust I have done
　my duty . . . but I own it has come upon me
　by surprise.
　Duke of York, on his deathbed, 1827

5　The noble Duke of York,
　He had ten thousand men,
　He marched them up to the top of the hill,
　And he marched them down again.
　Anon. song, 18th century

6　The best friend a soldier ever had.
　Hester Stanhope, *Memoirs* (1845)

7　It was he who had reduced chaos to order,
　restored discipline and, with discipline,
　confidence, had made the British Army the
　most efficient in the world.
　Sir John Fortescue, *History of the British
　Army*, XI (1923)

158　ZHUKOV
Georgi Konstantinovich, Marshal of the
Soviet Union (1895–1974)

1　If we come to a minefield, our infantry
　attacks exactly as it were not there.
　Marshal Georgi Konstantinovich Zhukov,
　remark to General Eisenhower in 1945,
　quoted in Salisbury, *Marshal Zhukov's
　Greatest Battles* (1969)

2　All the good people of the world who look
　back on those terrible days of the war when
　the fate of mankind hung in the balance will
　remember with respect and affection those
　who did not spare their lives fighting for the
　common cause, for the fate of their country,
　for the freedom and independence of all
　nations.
　Marshal Georgi Konstantinovich Zhukov,
　'The Battle of Berlin', *Voyenno-Istoricheski
　Zhurnal* (1965)

3　History shows that risks should be taken but
　not blindly.
　Marshal Georgi Konstantinovich Zhukov,
　'The Battle of Berlin', *Voyenno-Istoricheski
　Zhurnal* (1965)

4　All Soviet military units fought with
　extraordinary courage and valour in the
　fierce Battle for Moscow. Everyone, from
　private to general, displayed great heroism
　in fulfilling his sacred duty to the homeland,
　sparing neither his strength nor his very life
　in the defence of the city.
　Marshal Georgi Konstantinovich Zhukov,
　'The Battle for Moscow',
　Voyenno-Istoricheski Zhurnal (1966)

5　Bourgeois historians and former Nazi
　generals have tried to convince the public
　that the million picked German troops were
　beaten at Moscow not by the iron
　steadfastness, courage and heroism of Soviet
　soldiers, but by mud, cold and deep snow.
　The authors of these apologetics seem to
　forget that Soviet forces had to operate
　under the same conditions.
　Marshal Georgi Konstantinovich Zhukov,
　'The Battle for Moscow',
　Voyenno-Istoricheski Zhurnal (1966)

6　The working people of Moscow vowed to
　fight to the last with the soldiers rather than
　let the enemy through to the capital. And
　they kept that vow with honour.
　Marshal Georgi Konstantinovich Zhukov,
　'The Battle for Moscow',
　Voyenno-Istoricheski Zhurnal (1966)

7　It can truly be said that the front and the
　rear worked as one. Everyone did his utmost
　for victory over the enemy — irrefutable
　evidence of the common goals of our people
　and the armed forces in the struggle for their
　Socialist victory.
　Marshal Georgi Konstantinovich Zhukov,
　'The Battle of Kursk-Orel',
　Voyenno-Istoricheski Zhurnal (1967)

8　After Stalingrad no one really challenged
　Zhukov's primacy. His fellow marshals still
　competed with him for top honours. But he
　was Number One. And after Stalingrad no
　one really doubted that Russia with Zhukov
　at the head of her armies would finally
　defeat Germany.
　Harrison E. Salisbury, *Marshal Zhukov's
　Greatest Battles* (1969)

II
BATTLES AND WARS

159 BATTLE OF AGINCOURT 1415

1 The French fell in masses, the living and the dead piled helplessly together.
Enguerrand de Monstrelet, *Croniques* (1415), trs. Johnes

2 This day is called the feast of Crispian.
He that outlives this day, and comes safe home,
Will stand a tip-toe when this day is named,
And rouse him at the name of Crispian.
William Shakespeare, *Henry V*, VI, iii

3 We few, we happy few, we band of brothers;
For he today that sheds his blood with me
Shall be my brother; be he ne'er so vile
This day shall gentle his condition;
And gentlemen in England now a-bed
Shall think themselves accursed they were not here,
And hold their manhoods cheap whiles any speaks
That fought with us upon St Crispin's Day.
William Shakespeare, *Henry V*, IV, iii

4 Upon Saint Crispin's Day
Fought was this noble fray,
Which fame did not delay
To England to carry.
Michael Drayton, 'The Ballad of Agincourt' (1606)

5 The bare face of the ploughed field that had become a field of battle had grown a crop of arrow shafts as thick as a field of wind-laid wheat.
Weston Martyr, 'Bowmen's Battle', *Blackwood's Magazine*, 1938

6 Agincourt, like so many battles, was decided more by the stupidity of the loser than the brilliance of the victor.
Marjorie Ward, *The Blessed Trade* (1971)

160 BATTLE OF ALAMEIN 1942

1 The Tommies are bound to attack — for political reasons they've got no choice. But they are none too happy about it. We're going to wipe the floor with the British.
General Georg Stumme, Diary before the Battle of El Alamein, 1942

2 To your troops you can only offer one path — the path that leads to Victory or Death.
Adolf Hitler, signal to Field Marshal Erwin Rommel, 1942

3 I did not hope for such a complete victory; or rather I hoped for it but I did not expect it.
Field Marshal Viscount Montgomery, quoted in the *Daily Telegraph*, 1942

4 Montgomery has driven the election off America's front page.
'US News Front' column, *Daily Express*, 1942

5 It may almost be said, 'Before Alamein we never had a victory. After Alamein we never had a defeat.'
Winston S. Churchill, *The Second World War*, vol. IV (1951)

6   Alamein was a decisive victory but not a complete one.
Field Marshal Earl Alexander, *Memoirs* (1962)

7   The tide had finally and irrevocably turned. This was the best moment experienced by the British Army since another November day long ago in 1918.
General Sir David Fraser, *And We Shall Shock Them* (1983)

---

161 AMERICAN CIVIL WAR 1861–5
See also 181. GETTYSBURG

---

1   'Hurrah! Hurrah! we bring the Jubilee! Hurrah! Hurrah! the flag that makes you free.'
So we sang the chorus from Atlanta to the sea,
As we were marching through Georgia.
Henry C. Work, 'Marching through Georgia'

2   All quiet along the Potomac.
General George B. McClellan, attr. 1861

3   The destruction of the Navy at New Orleans was a sad, sad blow. . . .
Confederate Secretary Mallory after the fall of New Orleans on 24 April 1862

4   Of course we will get along together elegantly. All I have he can command, and I know the same feeling pervades every sailor's and soldier's heart. We are as one.
Major General William T. Sherman on the arrival of Lieutenant Commander S.L. Phelps for combined operations on the Tennessee River, 1863

5   Well, well, General, bury these poor men, and let us say no more about it.
General Robert E. Lee, Order to General A.P. Hill, after the Battle of Bristoe Station, 1863

6   In a large sense we cannot dedicate, we cannot hallow this ground. The brave men, living and dead, who struggled here, have consecrated it far above our power to add or detract.
Abraham Lincoln, address at Gettysburg, 1864

7   They couldn't hit an elephant at this distance . . .
General John Sedgwick, last words at Spottsylvania Court House, Virginia, 1864

8   Damn the torpedoes, full speed ahead!
Rear Admiral Farragut at Mobile Bay, 5 August 1864

9   I am prepared to sacrifice life, and will only surrender when I have no means of defense.
Brigadier General Richard L. Page refusing to surrender Fort Morgan despite overwhelming odds, 9 August 1864

10   Enough lives have been extinguished. We must extinguish our resentments if we expect harmony and union.
Abraham Lincoln, speech to his Cabinet, shortly before his assassination, 1865

11   To him our gratitude was justly due, for to him, under God, more than to any other person, we are indebted for the successful vindication of the integrity of the Union and the maintenance of the power of the Republic.
Secretary Welles announcing the assassination of President Lincoln, 1865

12   My opinion is that the Northern states will manage somehow to muddle through.
John Bright, quoted in McCarthy, *Reminiscences* (1899)

13   The war was ended only by the extinction of the military power of the South and its occupation.
Correlli Barnett, *Britain and Her Army* (1970)

14   The Civil War had been a war of the people, miscellaneously armed, shoddily clad, often unwillingly conscripted and corruptly officered, ragged in step, clumsy in drill, uncertain and sometimes panic-stricken in the face of the enemy — blue or grey.
John Keegan, *Six Armies in Normandy* (1982)

162 AMERICAN WAR OF INDEPENDENCE 1775–83

See also 169. BUNKER HILL

1 I know not what course others may take; but as for me, give me liberty; or give me death!
Patrick Henry, speech in Virginia House of Delegates, 1775

2 Stand your ground! Don't fire unless fired upon! But if you want to have a war, let it begin here!
Captain John Parker, Order to his Minutemen before the Battle of Lexington, 1775

3 If ever there was a just war since the world began, it is this in which America is now engaged.
Thomas Paine, *The Crisis* (1776)

4 If every nerve is not strained to recruit the new army with all proper expedience, I think the game is pretty near up.
General George Washington, letter to his brother, 1776

5 If I were an American, as I am an Englishman, while a foreign troop was landed in my country I would never lay down my arms — never, never, never!
William Pitt the Elder, speech in the House of Commons, 1777

6 The British officers in general behaved like boys who had been whipped at school. Some bit their lips, some pouted, others cried. Their round, broad-rimmed hats were well adapted to the occasion, hiding those faces they were ashamed to show.
Anon. American officer after the surrender of Yorktown, 1781, quoted in Wright, *The Fire of Liberty* (1984)

163 ASHANTI WAR 1873–4

1 Patronise beer, claret and good wines, wear a cholera belt about the loins, wear flannel next to the skin, bathe twice a day, avoid the sun, exercise moderately, but not in the early morning or at night.
Henry Morton Stanley, advice to member of Wolseley's British Expeditionary Force to the Gold Coast, in the *New York Herald*, 1873

2 Soldiers and sailors, remember the black man holds you in superstitious awe; fire low; fire slow, and charge home; the more numerous your enemy the greater will be the loss inflicted on him, and the greater your honour in defeating him.
Field Marshal Viscount Wolseley, Order of the Day to the British forces before Kumasi, 1874

3 Two thousand Ashantis, under the leadership of an intelligent British officer, would soon extend the power of the English from Cape Coast castle across the Thogoshi mountains to Timbuctoo, and from Mandingo land to Benin.
Henry Morton Stanley, *Coomassie and Magdala* (1874)

4 The most horrible war I ever took part in.
Field Marshal Viscount Wolseley, *The Story of a Soldier's Life*, vol. I (1903)

164 BATTLE OF AUSTERLITZ 1805

1 It is no dishonour to be defeated by *my* army.
Napoleon I, to a captured Russian officer after the Battle of Austerlitz, 1805

2 Soldiers, you are the first warriors of the world! Thousands of ages hence it will be told how a Russian army, hired by the gold of England, was annihilated by you on the plains of Olmütz.
Napoleon I, address to the French Army after their victory at the Battle of Austerlitz, 1805

3 Roll up that map; it will not be wanted these ten years.
William Pitt the Younger, on hearing the news of Napoleon's victory at the Battle of Austerlitz, 1805

165 BATTLE OF BANNOCKBURN 1314

1 (Of the English army) They glory in their warhorses and equipment. For us the name of the Lord must be our hope of victory in battle.
King Robert I of Scotland, address to the Scottish army before the Battle of Bannockburn, 1314, quoted in John of Fordun, *Chronica Gentis Scotorum* (*c.* 1383), ed. Skene

2 An evil, miserable and calamitous day for the English.
The Chronicle of Lanercost (1314), trs. Maxwell

3 (King Robert the Bruce's address to his army)
And when it cummis to the ficht,
Ilk man set his hert and micht
To stint our fais mekill pride.
John Barbour, *The Bruce* (*c.*1375)

4 (Of the rout of the English army)
Than mycht men her enseyneis cry:
And Scottish men cry hardely,
'On thaim! On thaim! On thaim! Thai faile!'
With sa hard thai gan assaile,
And slew all thai mycht our ta [all they could get at].
John Barbour, *The Bruce* (*c.*1375)

5 Maydens of Englonde, sore may yer morne
For your lemans ye have loste at Bannockisborne,
With heve a lowe.
What wenyth the kynge of Englonde
So soone to have wonne Scotlande
With rumbylowe.
Anon., quoted in Fabyan, *The Concordance of Chronicles* (1516)

166 BIBLICAL WARS

1 (Of the death of Sihon) And Israel smote him with the edge of the sword.
Numbers 21:24

2 (Of the Fall of Jericho) And it shall come to pass, that when they make a long blast with the ram's horn, and when ye hear the sound of the trumpet, all the people shall shout with a great shout; and the wall of the city shall fall down flat.
Joshua 6:5

3 (Of Gideon's deliverance) And he divided the three hundred men into three companies, and he put a trumpet in every man's hand, with empty pitchers, and lamps within the pitchers.
Judges 7:16

4 (Of Samson slaying the Philistines) And he smote them hip and thigh with a great slaughter.
Judges 15:8

5 (Of Samson slaying the Philistines) And he found a new jawbone of an ass, and he put forth his hand, and took it, and slew a thousand men therewith.
Judges 15:15

6 Saul hath slain his thousands, and David his ten thousands.
I Samuel 18:7

7 And the Philistines put themselves in array against Israel: and when they joined battle, Israel was smitten before the Philistines: and they slew of the army in the field about four thousand men.
I Samuel 4:2

8 (Of the slaying of Goliath) And David put his hand in his bag, and took thence a stone, and slang it, and smote the Philistine in his forehead, that the stone sunk in his forehead; and he fell upon his face to the earth.
I Samuel 17:49

9 How are the mighty fallen, and the weapons of war perished!
II Samuel 1:27

10 (Of the death of King Ahab) And one washed the chariot in the pool of Samaria; and the dogs licked up his blood; and they washed his armour; according unto the word of the Lord which he spake.
I Kings 22:38

11 Some trust in chariots, and some in horses: but we will remember the name of the Lord our God.
Psalms 20:7

167 BATTLE OF BLENHEIM 1704

1 I have not time to say more, but to beg you
will give my duty to the Queen, and to let
her know her army has had a glorious
victory.
Duke of Marlborough, letter to his wife,
1704

2 I have not a squadron or battalion which did
not charge four times at least.
Prince Eugene of Savoy, remark after the
battle, 1704, quoted in Coxe, *Memoirs of
the Duke of Marlborough* (1847)

3 'Twas a famous victory.
Robert Southey, 'The Battle of Blenheim'
(1798)

4 All Europe was hushed before these
prodigious events.
Winston S. Churchill, *A History of the
English Speaking Peoples,* vol. III (1957)

168 BOER WAR 1899–1903

1 A long and bloody war is before us, and the
end is by no means as certain as most people
imagine.
Winston S. Churchill, letter to Sir Evelyn
Wood, 1899

2 The Boer does not make a business of war,
and when he is not actually fighting he
pretends that he is camping out for pleasure.
Richard Harding Davis, *With Both Armies
in South Africa* (1900)

3 Far more people have been killed in our
hospitals than by Boer bullets.
Lady Edward Cecil, letter to the Marquis of
Salisbury from Bloemfontein, 1900

4 God seems to be with the Boers and against
us.
Field Marshal Viscount Wolseley, letter to
his wife, 1900

5 (Of the Relief of Kimberley) The people in
Kimberley looked fat and well. It was the
relieving force which needed food!
Field Marshal Earl Haig, letter to Colonel
Lonsdale Hale, 1900

6 The week which extended from December
10th to December 17th 1899 was the
blackest one known during our generation,
and the most disastrous for British arms
during this century.
Arthur Conan Doyle, *The Great Boer War*
(1900)

7 Who would have thought when you left
Johannesburg that I should be a *year* in
command with the war still going on?
Field Marshal Earl Kitchener, letter to Lord
Roberts describing his inability to bring the
Boer War to a successful conclusion, 1901

8 Let us admit it freely, as a business people
            should,
We have had no end of a lesson; it will do us
            no end of good.
Rudyard Kipling, 'The Lesson' (1903)

9 The Boers said the war was for liberty. The
British said it was for equality. The majority
of the inhabitants, who were not white at
all, gained neither liberty nor equality.
Rayne Kruger, *Goodbye Dolly Gray* (1959)

10 Whatever their idiosyncracies in terms of
dress or discipline, the Boers were, quite
simply, the best mounted infantrymen in the
world.
Richard Holmes, *The Little Field-Marshal*
(1981)

169 BATTLE OF BUNKER HILL 1775

1 You must drive these farmers from the hill
or it will be impossible for us to remain in
Boston.
General Sir William Howe, address to
British troops before the Battle of Bunker
Hill, 1775

2 The retreat was no flight; it was even
covered with bravery and military skill, and
proceeded no further than to the next hill,
where a new post was taken.

General John Burgoyne, after the Battle of Bunker Hill, 1775, quoted in Lloyd, *A Review of the History of Infantry* (1908)

3 The British had captured the hill, but the Americans had won the glory.
Winston S. Churchill, *A History of the English Speaking Peoples*, vol. II (1957)

---

170 CHARGE OF THE LIGHT BRIGADE 1854

---

1 There, my Lord! There is your enemy! There are your guns!
Captain Lewis Edward Nolan, retort to Lord Lucan before the Battle of Balaclava and the Charge of the Light Brigade, 1854

2 *C'est magnifique, mais ce n'est pas la guerre.*
It is magnificent, but it is not war.
Marshal Bosquet, on the Charge of the Light Brigade, 1854

3 The rumour in camp is that someone has been blundering, and that the Light Cavalry charge was all a mistake; the truth will come out some day.
Sergeant Thomas Gowing, letter to his parents from Sebastopol, 1854

4 A more fearful spectacle was never witnessed than by those who, without the power to aid, beheld their heroic countrymen rushing to the arms of death.
William Howard Russell, reporting the Charge of the Light Brigade at Balaclava in *The Times*, 1854

5 As far as it engendered excitement the finest run in Leicestershire could hardly bear comparison.
Lord George Paget, second-in-command of the Light Brigade, after the charge, 1854, quoted in his *The Light Cavalry Brigade in the Crimea* (1881)

6 When can their glory fade?
O the wild charge they made!
All the world wondered.
Honour the charge they made!
Honour the Light Brigade.
Noble six hundred.
Lord Tennyson, 'The Charge of the Light Brigade' (1854)

7 Six hundred stalwart warriors of England's pride the best,
Did grasp the lance and sabre on Balaclava's crest,
And with their trusty leader, Earl Cardigan, the brave,
Dashed through the Russian valley, to glory or a grave!
Slade Murray, 'Oh! 'Tis a Famous Story!' (1854)

---

171 CRIMEAN WAR 1854–6

See also 170. CHARGE OF THE LIGHT BRIGADE

---

1 For the peace, that I deem'd no peace, is over and done,
And now by the side of the Black and the Baltic deep,
And deathful-grinning mouths of the fortress, flames
The blood-red blossom of war with a heart of fire.
Lord Tennyson, *Maud* (1854)

2 Now France and Britain hand in hand,
What foe on earth could them withstand?
So let it run throughout the land,
The victory won at Alma.
Corporal John Brown, 'Battle of Alma' (1854)

3 Glory to each and to all, and the charge that they made!
Glory to all the three hundred, and all the Brigade.
Lord Tennyson, 'The Charge of the Heavy Brigade at Balaclava' (1854)

4 They dash on towards that thin red streak topped with a line of steel.
William Howard Russell, despatch describing the Russian Heavy Cavalry's charge against the 93rd Highlanders, *The Times*, 1854

5 I fear we are in a mess. We must try the bayonet.
General Sir Douglas Cathcart, last words at the Battle of Inkerman, 1854

6 We have no Wellington here.
Lt-Gen Sir Charles Windham, after the Battle of the Alma, Diary, 1854

7 We were going out to defend a rotten cause, a race that almost every Christian despises. However, as soldiers we had nothing to do with politics.
Sergeant Thomas Gowing, Diary, 1854, quoted in Fenwick, *Voices from the Ranks* (1954)

8 The sufferings of the army were unspeakable.
Justin McCarthy, *A History of Our Times* (1877)

9 The Crimean War is one of the two bad jokes in history.
Philip Guedalla, *The Two Marshals* (1943)

10 (Of the British Army in the Crimea) The conditions of service were intolerable; the administration was bad, the equipment scanty, the commanders of no outstanding ability.
Winston S. Churchill, *A History of the English Speaking Peoples*, vol. IV (1958)

11 The Crimean War is one of the compulsive subjects of British historical writing.
Correlli Barnett, *Britain and Her Army* (1970)

---

172 THE CRUSADES

---

1 On one side there will be poor wretches, on the other the truly rich; there the enemies of God, here his friends.
Pope Urban II, appeal to the Christian West, 1095, quoted in Fulcher of Chartres, *Gesta Francorum Iherusalem Peregrinantium* (1101–27)

2 What a sweet and wonderful sight it was for us to see all those shining crosses, whether of silk or gold or stuff, that at the Pope's order the pilgrims, as soon as they had sworn to go, sewed on the shoulders of their cloaks, their cassocks or their tunics!
Fulcher of Chartres, 1095, *Gesta Francorum Iherusalem Peregrinatium* (1101–27)

3 Our men advanced like clerks in a procession, and in truth it was a procession, for priests and many monks wearing white robes led the army of our soldiers, singing and calling on the help of God and the support of the saints.
Raymond of Aguilers, *Historia Francorum qui Ceperunt Jerusalem* (1099)

4 Who could enumerate this formidable army of Christ? No one, I think, had ever seen or could ever see a like number of such brilliant knights.
Anonymi Gesta Francorum (1101)

5 The Franks (may Allah forsake them!) have none of the better qualities of men, except courage.
Usama ibn Munqidh, *Autobiography* (1140)

6 (Of the attack on Damascus, 1148) Here then the fairest flowers of France faded before they had been able to bear fruit in the city of Damascus.
Odo of Deuil, *De Ludovici VII profectione in Orientum* (*c.* 1150)

7 On both sides this war was regarded as a religious matter.
Ibn al-Athir, of the fall of Jerusalem, 1187, quoted in *Recueils des Historiens des Croisades* (1872–1906)

8 The Christians were lions at the beginning of the fight, and at the end were no more than scattered sheep.
Imad ed-Din, of the Battle of Hattin, 1187, quoted in *Recueil des Historiens des Croisades* (1872–1906)

9 We shall die free or we shall triumph with glory.
Balian of Jerusalem, ultimatum to Saladin, 1187, quoted by Ibn al-Athir in *Recueils des Historiens des Croisades* (1872–1906)

10 As long as our enemies are hastening hither by sea and land, our country is threatened by the greatest disasters; and we see with amazement the pugnacity of the infidels and the indifference of the true believers.
Saladin, letter to the Caliph of Baghdad, 1189

11 (Of the Crusaders) Observe the steadfastness of this people, exposing themselves in this way to the most painful fatigues, without being paid or gaining any real advantage.
Beha al-Din, 1190, quoted in *Recueil des Historiens des Croisades* (1872–1906)

12 All those who shall take the cross and serve God for one year in the army shall be acquitted of all the sins they have sinned and have confessed.
Pope Innocent III, Indulgence, 1200

13 (Of the Fourth Crusade) Never had people undertaken so great a matter since the world was created.
Geoffrey of Villehardouin, *La Conquête de Constantinople* (1209)

---

173 BATTLE OF CULLODEN 1746

---

1 We are putting an end to a bad affair.
Lord George Murray, before the Battle of Culloden, 1746, quoted in Elcho, *A Short Account of the Affairs of Scotland* (1897)

2 All the good we have done is a little blood-letting, which has only weakened this madness, not cured it. I tremble for fear that this vile spot may still be the ruin of this island and our family.
Duke of Cumberland, letter to the Duke of Newcastle after the Battle of Culloden, 1746

3 Drumossie moor — Drumossie day —
A waefu' day it was to me!
For there I lost my father dear,
My father dear and brethren three.
Robert Burns, 'Drumossie Moor'

4 As the Highlanders were completely exhausted with hunger, fatigue and want of sleep, our defeat did not at all surprise me; I was only astonished to see them behave so well.
Chevalier James Johnstone, *Memoirs* (1820)

5 A lost cause will always win a last victory in men's imaginations.
John Prebble, *Culloden* (1961)

---

174 RETREAT FROM DUNKIRK 1940

---

1 So long as the English tongue survives, the word Dunkirk will be spoken with reverence. For in that harbor, in such a hell as never blazed on earth before, at the end of a last battle, the rags and blemishes that have hidden the soul of democracy fell away. There, beaten but unconquered, in shining splendour, she faced the enemy.
Editorial in the *New York Times*, 1940

2 We must be very careful not to assign to this deliverance the attributes of a victory. Wars are not won by evacuations.
Winston S. Churchill, speech in House of Commons, 1940

3 Germany's defeat and Europe's liberation began at Dunkirk.
Chester Wilmot, *The Struggle for Europe* (1952)

4 If Hitler had thrown the full weight of his armies into destroying the BEF, it could never have escaped.
Field Marshal Earl Alexander, *Memoirs* (1962)

5 Never was a great disaster more easily preventable.
Captain Sir Basil Liddell Hart, in the *Observer*, 1965

---

175 EASTER RISING 1916

---

1 The fools! The fools! The fools! They have left us our Fenian dead, and while Ireland holds these graves Ireland unfree shall never be at peace.
Patrick Pearse, speech at the funeral of Jeremiah O'Donovan Rossa, Dublin, 1915

2 In the name of God, and of the dead generations from which she receives her old tradition of nationhood Ireland through us summons her children to her flag and strikes for freedom.
Patrick Pearse, Proclamation at the General Post Office, Dublin, 1916

3 A terrible beauty is born.
W.B. Yeats, 'Easter 1916'

176 ENGLISH CIVIL WAR 1642–9

1 It was easy to begin the war, but no man knew where it would end.
Earl of Manchester, after the Battle of Marston Moor, 1644

2 We drove the entire cavalry of the Prince off the field. God made them as stubble to our swords.
Oliver Cromwell, after the defeat of the Royalist cavalry at the Battle of Marston Moor, 1644

3 No men undergo such hardship and hazard as the Souldier doth; none deserve better than they, either of Church, Commonwealth or posterity.
R. Ram, *The Souldier's Catechism* (1644)

4 Our victories seem to have been put in a bag with holes; what we won one time, we lost another. The treasure is exhausted; the country is wasted.
John Rushmore, address to Parliament, 1644

5 I can say this of Naseby, that when I saw the enemy draw up and march in gallant order towards us, and we a company of poor ignorant men to seek how to order our battle, the General having commanded me to order all the horse, I could not riding alone about my business, but smile to God in praises in assurance of victory, because God would, by things that are not, bring to naught things that are.
Oliver Cromwell, after the Battle of Naseby, 1645

6 Well, boys, you have done your work, and may go home and play — unless you fall out with one another.
Sir Jacob Astley, after surrendering to the Parliamentary Army at Stow-on-the-Wold, 1646

7 There are many thousands of us soldiers that have ventured our lives; we have had little propriety in the kingdom as to our estates, yet we have had a birthright. But it seems now except a man hath a fixed estate in this kingdom, he hath no right in this kingdom. I wonder we were so much deceived.
Edward Sexby, Putney Debates, 1647

8 A commonwealth army is like John the Baptist, who levels the mountains to the valleys, pulls down the tyrants, and lifts up the oppressed; and so makes way for the spirit of peace and freedom to come in and inherit the earth.
Gerrard Winstanley, *The Law of freedom, in A Platform* (1652)

9 Every county had the civil war, more or less, within itself.
Lucy Hutchinson, *Hutchinson Memoirs* (*c.*1670)

10 It was a deplorable thing that England should be at the mercy of a great body of soldiers with arms in their hands.
Charles Dickens, *A Child's History of England* (1852–4)

11 (Of Cromwell's victory) It was the triumph of some twenty thousand resolute, ruthless, disciplined, military fanatics over all that England has ever willed or ever wished.
Winston S. Churchill, *A History of the English Speaking Peoples*, vol. II (1956)

177 FALKLANDS WAR 1982

1 I don't mind about you, but I'm going down there to win the war.
Rear-Admiral John Woodward, after ordering the main British task force to sail from Ascension Island, 1982, quoted in *The Falklands War: The Full Story* (1982)

2 Rejoice, just rejoice!
Margaret Thatcher, after the retaking of South Georgia, 1982

3 This has been a pimple in the ass of progress festering for two hundred years, and I guess someone decided to lance it.
General Alexander Haig, quoted in the *Sunday Times*, 1982

4 It was like liberating an English suburban golf club.
Max Hastings, after the recapture of Port Stanley, in the *Standard*, 1982

5 All this crap about being educated from birth about the Malvinas. If they were that

committed, why didn't they fight for it?
Major Chris Keeble, after the defeat of the
Argentinian forces at Goose Green, 1982,
quoted in *The Falklands War: The Full Story*
(1982)

6 The British were going to war as they had
always gone, in haste and some confusion
but with confidence and great pride.
Max Hastings and Simon Jenkins, *The
Battle for the Falklands* (1983)

7 The Falklands came as a lifebelt to the navy.
Field Marshal Lord Carver, *Twentieth
Century Warriors* (1987)

8 When thousands of fighting troops suddenly
march into your house to tell you, with the
barrel of a gun stuck up your nose, that you
must no longer speak English, but Spanish,
you have a right to be defended by any
civilised nation.
Lt Robert Lawrence, *When the Fighting is
Over* (1988)

---

178 FIRST WORLD WAR 1914–18

See also 180. GALLIPOLI, 188. MARNE,
190. MONS, 193. PASSCHENDAELE,
202. SOMME

---

1 Wherever I look, to China, India, Egypt,
South Africa, Morocco and to Europe,
everything is restless and unsettled and
everyone except ourselves is getting ready
for war. This frightens me.
General Sir Henry Wilson, Diary, 1913

2 Just for a word — 'neutrality', a word which
in wartime has so often been disregarded,
just for a scrap of paper — Great Britain is
going to make war.
Theobald von Bethmann-Hollweg, letter to
Sir Edward Grey, 1914

3 The lamps are going out all over Europe; we
shall not see them lit again in our lifetime.
Sir Edward Grey, quoted from Grey,
Twenty-Five Years, 1892–1916, vol. II
(1925)

4 In Flanders' fields the poppies blow
Between the crosses, row on row.
John McCrae, 'In Flanders' Fields' (1914)

5 Keep the Home Fires Burning.
Lena Guilbert Ford, music-hall song, 'Till
the Boys Come Home' (1914)

6 When you go over the top you can slope
arms, light up your pipes and cigarettes, and
march all the way to Pozieres before meeting
any live Germans.
Brig-Gen H. Gordon, Order to troops of the
8th King's Own Yorkshire Light Infantry
before the Battle of the Somme, 1916

7 The world must be made safe for
democracy.
President Woodrow Wilson, address to
Congress, 1917

8 What is our task? To make Britain a fit
country for heroes to live in.
David Lloyd George, speech in
Wolverhampton, 1918

9 If any question why we died,
Tell them, because our fathers lied.
Rudyard Kipling, 'Common
Form' (1919)

10 The unfortunate thing is that this war has
become an overdose.
Captain Sir Basil Liddell Hart, 'Notes for
an Autobiography', 1920

11 My own attitude towards the conflict was
both simple and clear. In my eyes it was not
Austria fighting to get a little satisfaction
out of Serbia, but Germany fighting for her
life.
Adolf Hitler, *Mein Kampf* (1925)

12 Fifty years were spent in the process of
making Europe explosive. Five days were
enough to detonate it.
Captain Sir Basil Liddell Hart, *The Real
War, 1914–1918* (1930)

13 The line goes up and over the top,
Serious in gas masks, bayonets fixed,
Slowly forward — the swearing shells have
                    stopped —
Somewhere ahead of them death's
                    stopwatch ticks.
Peter Porter, 'Somme and Flanders',
Penguin Modern Poets (1962)

14 There is no doubt in my mind that the best
of England's men were killed in the war.
Captain C.S. Stormont Gibbs, *From the
Somme to the Armistice* (1986)

179 FRANCO–PRUSSIAN WAR 1870–1

1 Let there be but one party among you, that of France; one single flag, that of national honour. I am here in your midst. Faithful to my mission and to my duty, you will see me in the forefront of danger to defend the flag of France.
Empress Eugènie, Proclamation to the people of Paris after the defeat of the French Army at the Battle of Metz, 1870

2 We are in the chamber pot and there we shall be crapped upon.
General Ducrot, before the defeat of the French Army at the Battle of Sedan, 1870

3 Since I could not die in the midst of my troops, I can only put my sword in Your Majesty's hands. I am Your Majesty's good brother.
Emperor Louis-Napoleon of France, statement of surrender to Count Otto von Bismarck, 1870

4 Tied down and contained by the capital, the Prussians, far from home, anxious, harassed, hunted down by our re-awakened people, will be gradually decimated by our arms, by hunger, by natural causes.
Léon Gambetta, proclamation to the French partisans, Tours, 1870

5 What our sword has won in half a year, our sword must guard for half a century.
Helmuth, Graf von Moltke, speech after the Franco–Prussian War, 1871

6 The struggle of the working class against the capitalist class and its state has entered upon a new phase with the struggle in Paris. Whatever the immediate results may be, a new point of departure of world-historic importance had been gained.
Karl Marx, letter to Dr Kugelmann after the end of the Siege of Paris, 1871

7 Henceforth there are in Europe two nations which will be formidable, the one because it is victorious, the other because it is vanquished.
Victor Hugo, after the singing of the Treaty of Versailles, 1871

180 BATTLE OF GALLIPOLI 1915

1 I think it is folly to pour more troops and ammunition down the Dardanelles sink.
Field Marshal Earl Haig, letter to Colonel Lord Wigram, 1915

2 We were absolutely mad to embark on the infernal Dardanelles expedition.
Field Marshal Sir John French, letter to Winifred Bennett, 1915

3 Well, I have seen the place. It is an awful place and you will never get through.
Field Marshal Earl Kitchener, on visiting Gallipoli, 1915, quoted in Keyes, *Naval Memoirs,* vol. I (1934)

4 The long and varied annals of the British Army contain no more heart-breaking episode than the Battle of Suvla Bay.
Winston S. Churchill, *The World Crisis* (1915)

5 Perhaps as the years roll by we will be remembered as the expedition that was betrayed by jealousy, spite, indecision and treachery. The Turks did not beat us — we were beaten by our own High Command.
Private Joseph Murray, Diary, 1916, quoted in Laffin, *Damn the Dardanelles* (1980)

6 (Of the British and ANZAC troops) They went like kings in a pageant to the imminent death.
John Masefield, *Gallipoli* (1916)

7 I never heard anyone sing while I was at Anzac.
Douglas Hallam, 'Quinn's and Courtney's', *Blackwood's Magazine,* 1939

8 It will be up to date and probably to all eternity as sordid and miserable a chapter of amateur enterprise as ever was written in our history.
Lord Lovat, quoted in Kay, *Odyssey: Voices from Scotland's Recent Past* (1982)

181 BATTLE OF GETTYSBURG 1863

1 Never mind, General, all this has been my fault, it is I that have lost this fight, and you

must help me out of it the best way you can.
General Robert E. Lee, Order to General
C.M. Wilcox, after Pickett's charge, 1863

2   Then at the brief command of Lee
Moved out that matchless infantry,
With Pickett grandly leading down,
To rush against the roaring crown
Of those dread heights of destiny.
W.H. Thompson, *The High Tide at
Gettysburg* (1888)

3   Lee at Gettysburg no more than Napoleon
could win dominance.
Winston S. Churchill, *A History of the
English Speaking Peoples*, vol. IV (1958)

---

182 BATTLE OF HASTINGS 1066

---

1   We march straight on; we march to victory.
King Harold, rejecting the offer of a parley
before the Battle of Hastings, 1066

2   At last the English began to weary, and as if
confessing their crime in their defeat they
submitted to their punishment.
William of Poitiers, *Gesta Willielmi Ducis
Normannorum et Regis Anglorum* (*c.*1071)

3   A French bastard landing with an armed
banditti and establishing himself King of
England against the consent of the natives
is, in plain terms, a very paltry rascally
original.
Tom Paine, *Common Sense* (1776)

4   England itself, in foolish quarters of
England, still howls and execrates
lamentably over its William Conqueror, and
rigorous line of Normans and Plantagenets;
but without them, if you will consider it
well, what had it ever been?
Thomas Carlyle, *Frederick the Great*, I
(1858)

---

183 HUNDRED YEARS WAR 1337–1453

---

1   Now in the time of the noble Edward, who
has often put them to the test, the English

are the fiercest and most daring warriors
known to man.
Jean le Bel, *Chroniques*, I (1272)

2   It is a very good sign for me, for the land
wants me.
King Edward III of England, on falling
while landing at Hogue-Saint-Vast, 1346,
quoted in Froissart, *Chronicles* (1523–5)

3   No peace until they give back Calais.
Eustace Deschamps, 'Calais' (1387)

4   When I see that they [the English] want to
do nothing save lay waste and destroy this
realm, from may God preserve it, and how
they wage war to the death on all their
neighbours, I hold them in such
abomination and hatred that I love those
who hate them and hate those who love
them.
Jean de Montreuil, *A tout la chevalrie de
France* (1411)

5   The people, unnerved by a long period of
peace and order, simple as they were,
generally thought that the English were not
men like everyone else but wild beasts,
gigantic and ferocious, who were going to
throw themselves on them and devour them.
Thomas Basin, report to King Louis XI on
the poverty brought about by the war in
Normandy, 1417

6   This storm of war raised up against us by
the people of England.
Jean Chartier, *Chronique de Charles VII*
(*c.*1420)

7   (Of the meeting between King Edward IV
and King Louis XI, 1475) The English men,
after their wonted manner, first demanded
the Crown, at the least Normandy and
Guienne, but they were no more earnestly
demanded than stoutly denied.
Philip de Comines, *Memoirs* (1614), trs.
Dassett

8   All the time of war during those forty
betwixt England and France, wist I not
scant three or four men which wolden
accord throughout, in telling how a town or
castle was won in France, or how a battle
was done there.
Bishop Reginald Peacock, *c.* 1498, quoted
in Commynes, *Mémoires* (1498)

9   This astounding victory of Crécy ranks with
Blenheim, Waterloo and the final advance in

the last summer of the Great War as one of the four supreme achievements of the British Army.
Winston S. Churchill, written in 1939 and quoted in *A History of the English Speaking Peoples*, vol. I (1956)

184 INDIAN MUTINY 1857

1 My lot is cast for Lucknow. The enterprise of crossing the Ganges, opposed by double my numbers, is not without hazard. But it has, to me, at sixty-three, all the charm of a romance. I am as happy as a duck in thunder.
Brig-Gen Sir Henry Havelock, quoted in Marshman, *Memoirs of Sir Henry Havelock* (1860)

2 Hold on, and do not negotiate, but rather perish sword in hand.
Brig-Gen Sir Henry Havelock, despatch to Colonel John Inglis, during the Siege of Lucknow, 1857

3 I can never spare a sepoy again. All that fall into my hands will be dead men.
General James Neill, Order after finding the massacred bodies at Cawnpore, 1857

4 Let us propose a Bill for the flaying alive, impalement or burning of the murderers of the women and children at Delhi. The idea of simply hanging the perpetrators of such atrocities is maddening.
General John Nicholson, letter to Sir Herbert Edwardes, 1857

5 Some thousands that left England where
    first they drew their breath;
How sad to tell, in battle fell, and yielded
    unto death,
At Lucknow and at Delhi and taking of
    Cawnpore;
For years to come will history relate the
    India War.
Anon. song of 1858, 'The Great India War'

6 (Of the massacre at Cawnpore) There was not a soldier in the garrison who did not recoil from the thought of surrender — who would not have died with sword or musket

in hand rather than lay down his arms at the feet of the treacherous Mahratta.
John William Kaye, *The Sepoy War in India* (1876)

7 But sweetest of all music
The pipes at Lucknow played.
John Greenleaf Whittier, 'The Pipes at Lucknow', *In War Time* (1860)

8 Saved by the valour of Havelock, saved by
    the blessing of Heaven!
'Hold it for fifteen days!' We have held it for
    eighty-seven!
Lord Tennyson, 'The Defence of Lucknow' (1879)

9 When Mehtab Singh rode from the gate
His chin was on his breast:
The captains said, 'When the strong
    command
Obedience is the best.'
Sir Henry Newbolt, 'A Ballad of John Nicholson', *Poems New and Old* (1912)

185 KOREAN WAR 1950–3

1 The aggression of the North Koreans was, and remains, an international crime of the first order.
Michael Foot, writing in *Tribune*, 1950

2 We shall land at Inchon and I shall crush them.
General Douglas MacArthur, speech to the Joint Chiefs of Staff, Tokyo, 1950

3 Here we fight Europe's war with arms while the diplomats there still fight it with words.
General Douglas MacArthur, letter to Representative Joseph Martin, 1951

4 It was clear that there was something profoundly disturbing about this campaign and something profoundly disturbing about its Commander-in-Chief.
Reginald Thompson, *Cry Korea* (1951)

5 This Korean War is a Truman War.
Robert A. Taft, speech at Milwaukee, 1951

6 Korea has been a blessing. There had to be a Korea either here or some place in the world.

General James A. Van Fleet, 1952, quoted
in Aronson, *The Press and The Cold War*
(1970)

7 That job requires a personal trip to Korea. I
shall make that trip. Only in that way could
I learn best how to serve the American
people in the cause of peace. I shall go to
Korea.
General Dwight D. Eisenhower, speech in
Detroit during the presidential election
campaign, 1952

8 The Korean War began in a way in which
wars often begin. A potential aggressor
miscalculated.
John Foster Dulles, speech at St Louis, 1953

9 From the beginning there was the certainty
that this was a doomed place, a country of
despair; one was afflicted by both the
discomfort and that kind of tenderness
which, if you love Asia, means revulsion —
an emotion incapable of explanation or
analysis. I loathed Korea, and to this day it
lives perversely in my mind as though I
loved it.
James Cameron, *Point of Departure* (1967)

10 Most men's first impression of Korea was of
the stench, drifting out from the land to the
sea: of human excrement and unidentifiable
oriental exotica, mostly disagreeable.
Max Hastings, *The Korean War* (1987)

11 All those officers, those generals: they really
thought they were going to go over there
and 'stop the gooks' — just the same as in
Vietnam. Just who 'the gooks' were, they
didn't know, and didn't want to know.
Colonel Frank Ladd, quoted in Hastings,
The Korean War (1987)

---

186 BATTLE OF MALPLAQUET 1709

1 I am so tired that I have but strength to tell
you that we have had this day a very bloody
battle; the first part of the day we beat their
foot, and afterwards their horse.
Duke of Marlborough, letter to his wife,
1709

2 It was the most deliberate, solemn and
well-ordered battle that I ever saw.

Lt-Col John Blackader, Diary,
1709

3 Europe was appalled at the slaughter of
Malplaquet.
Winston S. Churchill, *A History of the
English Speaking Peoples*, vol. III (1957)

---

187 BATTLE OF MARATHON 490 BC

1 Fighting in the forefront of the Greeks, the
Athenians crushed at Marathon, the might
of the gold-bearing Medes.
Simonides, *Fragment*, 88

2 In that fight we stood alone against Persia
— we dared a mighty enterprise and came
out of it alive — we beat forty-six nations to
their knees!
Herodotus, *Histories* IX, 27, trs. de
Selincourt

3 That man is little to be envied whose
patriotism would not gain force upon the
plain of Marathon.
Samuel Johnson, *Journey to the Western
Isles of Scotland* (1775)

4 The mountains look on Marathon —
And Marathon looks on the sea;
And musing there an hour alone,
I dream'd that Greece might still be free.
Lord Byron, *Don Juan*, III (1821)

5 Truth-loving Persians do not dwell upon
The trivial skirmish fought near Marathon.
Robert Graves, 'The Persian Version',
Collected Poems (1975)

---

188 BATTLE OF THE MARNE 1914

1 Victory will come to the side that outlasts
the other.
Marshal Ferdinand Foch, Order to the
French Army during the Battle of the
Marne, 1914

2 The empire of blood and iron rolled slowly
back towards the darkness of the northern
forests; and the great nations of the West

went forward; where side by side as after a long lovers' quarrel, went the ensigns of St Denys and St George.
G.K. Chesterton, *The Crimes of England* (1915)

3 I don't know who won the Battle of the Marne, but if it had been lost, I know who would have lost it.
Marshal Joseph Joffre, evidence to the Briey Commission, 1919

4 Whatever the mistakes of the opening phase, however wrong the tactical and strategic conceptions which had induced them, immortal glory crowns the brows of those who gave the fateful signal, and lights the bayonets of the heroic armies that obeyed it.
Winston S. Churchill, *The World Crisis* (1923)

---

189 BATTLE OF MINDEN 1759

---

1 I have seen what I never thought to be possible, a single line of infantry break through three lines of cavalry ranked in order of battle and tumble them to ruin.
Marshal Contades, on the advance of the British and Hanoverian infantry at Minden, 1759, quoted in Lloyd, *A Review of the History of Infantry* (1908)

2 Never were so many boots and saddles seen on a battlefield as opposite to the English and Hanoverian Guards.
General Westphalen, after the Battle of Minden, 1759, quoted in Fortescue, *History of the British Army,* II (1910)

3 All preconcerted arrangements were upset by the extraordinary attack of the British infantry, a feat of gallantry and endurance that stands, so far as I know, absolutely without parallel.
Sir John Fortescue, *A History of the British Army,* II (1910)

---

190 BATTLE OF MONS 1914

---

1 Tramp, tramp, the grim road, the road from Mons to Wipers

(I've 'ammered out this ditty with me bruised and bleedin' feet.)
Tramp, tramp, the dim road — we didn't 'ave no pipers,
And bellies that was 'oller was the drums we 'ad to beat.
Robert Service, 'The Red Retreat', *Rhymes of a Red Cross Man* (1916)

2 What you did resulted in ultimate victory. History will record this to your glory.
Viscount Haldane, letter to Field Marshal Sir John French, 1919

3 It expressed in blood Britain's inevitable yet so long evaded commitment to the continent.
Correlli Barnett, *The Swordbearers* (1963)

4 Set beside what was happening elsewhere along the front, and what the Army was to experience later, Mons scarcely rates as a battle at all; there is certainly no evidence that it slowed the Germans to any noticeable extent.
John Terraine, *The Smoke and the Fire* (1980)

NAPOLEONIC WARS
See 164. AUSTERLITZ, 195. PENINSULAR, 210. WATERLOO

---

191 NUCLEAR WAR

---

1 The atomic bomb is not an inhuman weapon.
Maj-Gen Leslie R. Groves, public statement on the Manhattan Project, 1945

2 (Of the bombing of Hiroshima) Yesterday we clinched victory in the Pacific, but we sowed the whirlwind.
Hanson Baldwin, in the *New York Times,* 1945

3 Where both sides possess atomic power 'total warfare' makes nonsense.
Captain Sir Basil Liddell Hart, *The Revolution in Warfare* (1946)

4 By carrying destructiveness to a suicidal extreme, atomic power is stimulating and

accelerating a reversion to the indirect methods that are the essence of strategy — since they endow war with intelligent properties that raise it above the brute application of force.
Captain Sir Basil Liddell Hart, *Strategy* (1954)

5 Our capacity to retaliate must be, and is, massive in order to deter all forms of aggression.
John Foster Dulles, speech in Chicago, 1955

6 Mass will suffer most from weapons of mass destruction.
K.J. Macksey, *Land Warfare of the Future* (1956)

7 (Of nuclear weapons) The whole thing has a gruesome and horrific effect which makes one really fear for the sanity of mankind.
Admiral Earl Mountbatten, Diary, 1959

8 Every man, woman and child lives under a nuclear sword of Damocles, hanging by the slenderest of threads, capable of being cut at any moment by accident or miscalculation or madness.
John F. Kennedy, speech to United Nations Assembly, 1961

9 The idea of weapons of mass extermination is utterly horrible and is something which no one with one spark of humanity can tolerate.
Bertrand Russell, speech at Birmingham, 1961

10 We may expect in due course any nation large or small will need its H-bombs as a *nouveau riche* needs a grouse-moor and his wife a mink coat to establish status.
Compton Mackenzie, *On Moral Courage* (1962)

11 We have genuflected before the god of science only to find that it has given us the atomic bomb, producing fears and anxieties that science can never mitigate.
Martin Luther King, *Strength to Love* (1963)

12 Make no mistake. There is no such thing as a conventional nuclear weapon.
Lyndon B. Johnson, speech at Syracuse, New York, 1964

13 Whether the crowds are screaming for blood or waving banners of peace no longer seems

to matter when it needs only one man's finger to press the button.
Marjorie Ward, *The Blessed Trade* (1971)

192 BATTLE OF OMDURMAN 1898

1 After the fight all of us said it was the finest sight we should ever see in our lives.
Lieutenant H.V. Fison, letter to his father, 1898, quoted in Emery, *Marching Over Africa* (1986)

2 I shall merely say that the victory of Omdurman was disgraced by the inhuman slaughter of the wounded and that Kitchener was responsible for this.
Winston S. Churchill, letter to his mother, 1898

3 I think we've given them a good dusting, gentlemen.
Field Marshal Earl Kitchener, at the conclusion of the Battle of Omdurman, 1898

4 (Of the Dervish charge) They came very fast, they came very straight, and they presently came no further.
G.W. Steevens, *With Kitchener to Khartoum* (1898)

5 Line after seemingly irresistible line of dervishes made their way forward in tight battle order, a wave of humanity pressing on to break on the concentrated fire of the British and Egyptian infantry.
Trevor Royle, *Death Before Dishonour: The True Story of Fighting Mac* (1982)

193 BATTLE OF PASSCHENDAELE 1917

1 Did we really send men to fight in this?
Lt-Gen Sir Lancelot Kiggell, on visiting the battlefield, 1917

2 It was not the enemy but mud that prevented us doing better.
Field Marshal Earl Haig, Diary, 1917

3 Maybe one day we'll forget the rain
The mud and filth of a Belgian scene
But always in mem'ry I'll see again

Those roads with the stumps where the trees
    had been.
Anon., poem in *The Wipers Times*, 1917

4 The rain drives on, the stinking mud
becomes more evilly yellow, the shell-holes
fill up with green-white water, the roads and
tracks are covered in inches of slime, the
black dying trees ooze and sweat and the
shells never cease.
Paul Nash, letter to his wife from Ypres,
1917

5 What the German soldier experienced,
achieved, and suffered in the Flanders battle
will be his everlasting monument of bronze,
erected by himself in the enemy's land.
General Erich von Ludendorff, *My War
Memories* (1919)

6 He who was present can be proud to have
been a Flanders warrior.
Crown Prince Ruprecht of Bavaria, *In
Treue Fest* (1922)

7 We could not explain the Passchendaele
offensive from a military point of view,
consequently we could only see it as
political, a calm cold-blooded sacrifice of
life for miserable diplomatic advancement.
Hugh Quigley, *Passchendaele and the
Somme* (1928)

8 Third Ypres was the blindest slaughter of a
blind war.
A.J.P. Taylor, *The First World War* (1963)

---

194 PELOPONNESIAN WARS
459–446 BC, 431–404 BC

---

1 War is not so much a matter of armaments
as of the money which makes armaments
effective: particularly is this true in a war
fought between a land power and a sea
power.
King Archidamus, speech to the Spartans
before the declaration of war, 432 BC,
quoted in Thucydides, *Peloponnesian Wars*,
I, 83, trs. Warner

2 Remember that what makes a good soldier
is his readiness to fight, his sense of honour
and his discipline, and that this day, if you

show yourselves men, will win you your
freedom and the title of allies of Sparta.
Brasidas, speech to the Dorian troops before
the Battle of Amphipolis, 422 BC, quoted in
Thucydides, *Peloponnesian Wars*, V, 9, trs.
Warner

3 It is men who make the city and not walls or
ships with no men inside them.
Nicias, speech to the Athenian army in
Sicily, quoted in Thucydides, *Peloponnesian
Wars*, VII, 77, trs. Warner

4 Thucydides the Athenian wrote the history
of the war fought between Athens and
Sparta, beginning the account at the very
outbreak of the war, in the belief that it was
going to be a great war and more worth
writing about than any of those which had
taken place in the past.
Thucydides, *Peloponnesian Wars*, I, 1, trs.
Warner

5 As for this present war, even though people
are apt to think that the war in which they
are fighting is the greatest of all wars, and,
when it is over, to relapse again into their
admiration of the past, nevertheless, if one
looks at the facts themselves, one will see
that this was the greatest war of all.
Thucydides, *Peloponnesian Wars*, I, 21, trs.
Warner

6 What made war inevitable was the growth
of Athenian power and the fear which this
caused in Sparta.
Thucydides, *Peloponnesian Wars*, I, 23, trs.
Warner

7 In times of peace and prosperity cities and
individuals alike follow higher standards
because they are not forced into a situation
where they have to do what they do not
want to do. But war is a stern teacher; in
depriving them of the power of easily
satisfying their daily wants, it brings most
people's minds down to the level of their
actual circumstances.
Thucydides, *Peloponnesian Wars*, III, 82,
trs. Warner

8 I myself remember that all the time from the
beginning to the end of the war it was being
put about by many people that the war
would last thrice nine years.
Thucydides, *Peloponnesian Wars*, V, 26, trs.
Warner

195 PENINSULAR WAR 1808–13

1 I suspect that all the Continental armies
   were more than half beaten before the battle
   was begun. I, at least, will not be frightened
   beforehand.
   Duke of Wellington, statement before
   taking command, 1808, quoted in Croker,
   The Croker Papers (1884)

2 They are a fine people. The road is swarmed
   with armed peasants; every man would be a
   soldier. If Spain is subdued it will not be the
   fault of her people.
   Sir George Jackson, before Moore's retreat
   to Corunna, Diary, 1808

3 We are French, we are still breathing, and
   we are not the victors.
   Maurice de Tascher, after the French defeat
   at the Battle of Baylen, Diary, 1808

4 (Of the retreat to Corunna) The army may
   rest assured that there is nothing he has
   more at heart than their honour — and that
   of their Country.
   General Sir John Moore, Order at Corunna,
   1808

5 The people here have the cool effrontery to
   look upon the English troops as exotic
   animals who have come to engage in a
   private fight with the French and now that
   they are here all that the fine Spanish
   gentlemen have to do is to look on with
   their hands in their pockets.
   A.L.F. Schaumann, before the Battle of
   Corunna, 1808, quoted in *On the Road with
   Wellington* (1924) trs. Ludovici

6 The battle of Talavera was certainly the
   hardest fought of modern days, and the
   most glorious in its results to our troops.
   Duke of Wellington, letter to Colonel John
   Malcolm, 1809

7 (Of the Battle of Albuera, 1811) Then was
   seen with what a strength and majesty the
   British soldier fights.
   Maj-Gen Sir William Napier, *Peninsular
   War*, vol. XII (1886)

8 We lived united as men always are who are
   daily staring death in the face on the same
   side and who, caring little about it, look
   upon each new day added to their lives as
   one more to rejoice in.

   John Kincaid, of the British Army in 1811,
   quoted in *Random Shots from a Rifleman*
   (1835)

9 (Of the Battle of Salamanca, 1812) There
   was no mistake; everything went in as it
   ought; and there was never an army
   defeated in so short a time.
   Duke of Wellington, letter to Earl Bathurst,
   1812

10 Spain was before us and every man in the
   Rifles seemed only too anxious to get a rap
   at the French again. It was a glorious sight
   to see our colours spread in those fields. The
   men seemed invincible and nothing, I
   thought, could beat them.
   Rifleman John Harris, *Recollections* (1848)

11 They are the shipped battalions sent
   To bar the bold Belligerent
   Who stalks the Dancers' Land.
   Within these hulls, like sheep a-pen,
   Are packed in thousands fighting-men
   And colonels in command.
   Thomas Hardy, *The Dynasts*, II, ii
   (1904–08)

12 We fought right on with conqu'ring banners
                                    o'er us
   From Torres Vedras to the Pyrennees!
   Edward Fraser, 'Soldier's Song', *The
   Soldiers Whom Wellington Led* (1913)

196 PERSIAN WARS 490–479 BC
See also 187. MARATHON,
205. THERMOPYLAE

1 Haste is the mother of failure — and for
   failure we always pay a heavy price.
   Artabanus, warning to King Xerxes, quoted
   in Herodotus, *Histories*, VII, 10, trs. de
   Selincourt

2 No one is fool enough to choose war instead
   of peace — in peace sons bury fathers but in
   war fathers bury sons.
   Croesus, after being reprieved by King
   Cyrus, quoted in Herodotus, *Histories*, I,
   87, trs. de Selincourt

3 Force is always beside the point when
   subtlety will serve.

King Darius, Order to kill Oroetus, quoted in Herodotus, *Histories*, III, 127, trs. de Selincourt

4 Divine Salamis, you will bring death to
           women's sons
When the corn is scattered, or the harvest
           gathered in.
Priestess Aristonice, prophecy of the Oracle at Delphi before the invasion of Greece, 480 BC, quoted in Herodotus, *Histories*, VII, 143, trs. de Selincourt

5 Herodotus of Halicarnassus, his *Researches* are here set down to preserve the memory of the past by putting on record the astonishing achievements both of our own and other peoples; and more particularly to show how they came into conflict.
Herodotus, *Histories*, I, 1, trs. de Selincourt

6 It was the Athenians who — after God — drove back the Persian king.
Herodotus, *Histories*, VII, 139, trs. de Selincourt

7 The evil of internal strife is worse than united war in the same proportion as war itself is worse than peace.
Herodotus, *Histories*, VIII, 3, trs. de Selincourt

8 The worst pain a man can have is to know much and be impotent to act.
Herodotus, *Histories*, IX, 16, trs. de Selincourt

---

197 BATTLE OF PLASSEY 1757

1 We have lost our glory, honour and reputation, everywhere but in India.
William Pitt the Elder, announcing the news of the British victory at Plassey in the House of Commons, 1758

2 With the loss of twenty-two soldiers killed and fifty wounded, Clive had scattered an enemy of nearly sixty thousand men, and subdued an empire larger and more populous than Great Britain.
Lord Macaulay, *Life and Works*, vol. VI (1897)

3 The most miserable skirmish ever to be called a decisive battle.
Philip Mason, *The Men Who Ruled India*, vol. I (1953)

---

198 PUNIC WARS 264–241 BC, 218–201 BC, 149–146 BC

1 *Delenda est Carthago.* Carthage must be destroyed.
Cato the Elder, beginning of his speeches to the Senate during the Punic Wars, quoted in Plutarch, *Lives*, 8

2 We have accomplished nothing till we have stormed the gates of Rome, till our Carthaginian standard is set in the city's heart.
Hannibal, declaration of war, 218 BC, quoted in Juvenal, *Satires*, X, trs. Green

3 The Romans boldly aspired to universal dominion and, what is more, achieved what they aimed at.
Polybius, *Histories*, II, 38

4 (Of the Battle of Cannae, 216 BC) Neither the defeats they had suffered nor the subsequent defection of all these allied peoples moved the Romans ever to breathe a word about peace.
Livy, *Histories* XXII, 61

5 (Of the Second Punic War) This war, a struggle between the two wealthiest peoples in the world, had attracted the attention of all kings and all nations elsewhere.
Livy, *Histories*, XXIII, 33

6   That long war
Whose spoil was heaped so high with rings
          of gold.
Dante, *Divina Commedia, Inferno*, XXVIII, 10 (*c*.1315)

7 (Of the Battle of Cannae, 216 BC) If for one moment panic had unnerved the iron courage of the Roman aristocracy, on the next their inborn spirit revived; and their resolute will striving beyond its present power created, as in the law of our nature, the power which it required.
Thomas Arnold, *The Second Punic War* (1886)

---

199 RUSSO–JAPANESE WAR 1904–5

1 What this country needs is a short victorious war to stem the tide of revolution.

Vyacheslav Konstantinovich Plehve, 1903, quoted in Witte, *Reminiscences* (1922)

2 March then with our sunlight banner
Waving proudly in the van,
March beneath that glorious emblem,
Down with Russia! On Japan!
Lt-Gen Sir Ian Hamilton, translation of Japanese marching song, 1904

3 Warfare reverted to a primaeval scrimmage where brute bravery and fox-like cunning on the part of individuals or groups won the day.
Ashmead Bartlett, *The Times*, 1905

4 (Of the battle of Liaoyang, 1905) I have today seen the most stupendous spectacle it is possible for the mortal brain to conceive — Asia advancing, Europe falling back, the wall of mist and the writing thereon.
Lt-Gen Sir Ian Hamilton, *A Staff Officer's Scrap Book during the Russo–Japanese War* (1907)

---

200 SECOND WORLD WAR 1939–45

See also 160. ALAMEIN, 174. DUNKIRK, 204. STALINGRAD, 207. TOBRUK

---

1 I believe it is peace in our time.
Neville Chamberlain, radio broadcast after the Munich Agreement, 1938

2 The *Daily Express* declares that Britain will not be involved in a European war this year, or next year either.
Daily Express, front page, 1938

3 *C'est une drôle de querre*. It's a phoney war.
Edouard Daladier, speech in French Chamber of Deputies 1939

4 We shall not capitulate — no, never! We may be destroyed, but if we are, we shall drag a world with us — a world in flames.
Adolf Hitler, address at Nuremburg, 1939

5 The Navy can lose the war, but only the Air Force can win it. Therefore, our supreme effort must be to gain overwhelming mastery in the air.
Winston S. Churchill, to the War Cabinet, 1940

6 We are waiting for the long-promised invasion. So are the fishes.
Winston S. Churchill, radio broadcast to the French people, 1940

7 I'm glad we've been bombed. It makes me feel I can look the East End in the face.
Queen Elizabeth (now Queen Mother) after the bombing of Buckingham Palace, 1940

8 In occupied regions conditions must be made unbearable for the enemy and all his accomplices. They must be hounded and annihilated at every step and all their measures frustrated.
Josef Stalin, address to the Russian people ordering a scorched earth policy, 1941

9 The power of Germany must be broken in the battlefields of Europe.
Franklin D. Roosevelt, 1941, quoted in Bryant, *The Turn of the Tide* (1957)

10 This land was made for War.
Jocelyn Brooke, 'Landscape Near Tobruk'

11 We shall have failed, and the blood of our dearest will have flowed in vain, if the victory they died to win does not lead to a lasting peace, founded on justice and good will.
King George VI, Victory broadcast to the people of Britain, 1945

12 There were own, there were the others.
Their deaths were like their lives, human
and animal.
There were no gods and precious few
heroes.
Hamish Henderson, 'End of a Campaign', *Elegies for the Dead in Cyrenaica* (1948)

13 The two most serious miscalculations of the Second World War both concerned the Soviet Union: Hitler's miscalculations of Russia's military strength and Roosevelt's miscalculation of Russia's political ambition.
Chester Wilmot, *The Struggle for Europe* (1952)

14 The fighting in North Africa was hard on both sides, but it was fair.
Maj-Gen Friedrich Wilhelm von Mellinthin, *Panzer Battles* (1955)

15 The great immorality open to us in 1940 and 1941 was to lose the war against Hitler's Germany.
Dr Noble Frankland, Lecture to Royal United Service Institution, 1961

16 It was the *Eastern* Front which fulfilled the function of the Western Front in the Great War.
Correlli Barnett, *Britain and Her Army* (1970)

17 My feelings about the war as such were and are highly conventional, that it was very unpleasant, but altogether necessary; because the Germans had to be put down.
Kingsley Amis, quoted in Hamilton, *The Poetry of War, 1939–1945* (1972)

18 Despite all the killing and the destruction that accompanied it, the Second World War was a good war.
A.J.P. Taylor, *The Second World War* (1975)

19 In private the British and American commanders despised each other for lack of drive, attacking in insufficient strength and lack of appetite for combat.
Dominick Graham and Shelford Bidwell, *Tug of War: The Battle for Italy 1943–1945* (1986)

---

201 SEVEN YEARS WAR 1756–63
See also 189. MINDEN, 197. PLASSEY

---

1 I know that I can save this country and that no one else can.
William Pitt the Elder, to the Duke of Devonshire, on the outbreak of war, 1756

2 Our bells are going to answer the guns, which just now speak their joy.
Hester Pitt, Lady Chatham, letter to her husband on receiving the news of the victory at Ticonderoga, 1759

3 Our bells are worn threadbare with the ringing of victories.
Horace Walpole, letter to Edward Montagu in praise of the 'Year of Victories', 1759

4 Is this country to wage eternal war upon wild imaginary schemes of conquest?
Earl of Hardwicke, speech in the House of Lords, 1760

---

202 BATTLE OF THE SOMME 1916

---

1 If we had heroes in previous wars, today we have them not in thousands but in half millions.
Captain Reginald Leetham, Diary, 1916, quoted in Brown, *Tommy Goes to War* (1978)

2 It was an amazing spectacle of unexampled gallantry, courage and bull-dog determination on both sides.
Matthäus Gerster, *Die Schwaben an der Assare* (1918)

3 A sense of the inevitable broods over the battlefields of the Somme. The British armies were so ardent, their leaders so confident, the need and appeals of our Allies so clamant, and decisive results seemingly so near, that no human power could have prevented the attempt.
Winston S. Churchill, *The World Crisis 1916–1918* (1927)

4 The Somme battle raised the morale of the British Army. Although we did not win a decisive victory, there was what matters most, a definite and growing sense of superiority over the enemy, man to man.
Charles Carrington, *A Subaltern's War* (1929)

5 It proved both the glory and the graveyard of Kitchener's army.
Captain Sir Basil Liddell Hart, *The Real War 1914–1918* (1930)

6 The tragedy of the Somme battle was that the best soldiers, the stout-hearted men were lost; their numbers were replaceable, their spiritual worth could never be.
Anon. German soldier, quoted in Moran, *The Anatomy of Courage* (1945)

7 Idealism perished on the Somme.
A.J.P. Taylor, *The First World War* (1963)

8 Whatever was gained, it wasn't worth the price that the men had paid to gain that advantage. It was no advantage to anybody. It was just sheer bloody murder. That's the only words you can use for it.
Corporal Henry Shaw, quoted in Macdonald, *Somme* (1983)

---

203 SPANISH CIVIL WAR 1936–9

---

1 It is better to die on your feet than to live on your knees.
Dolores Ibarruri ('La Pasionara'), speech in Paris, 1936

2   *No pasarán!* They shall not pass.
    Anon., watchword during the defence of
    Madrid, 1936

3   Freedom is an easily spoken word
    But facts are stubborn things. Here, too, in
              Spain
    Our fight's not won till the workers of all
              the world
    Stand by our guard on Huesca's plain
    Swear that our dead fought not in vain,
    Raise the red flag triumphantly
    For Communism and for liberty.
    John Cornford, 'Full Moon at Tierz'
    (*c.*1936)

4   We shall not forget you, and, when the olive
    tree of peace puts forth its leaves again,
    mingled with the laurels of the Spanish
    people's victory — come back!
    Dolores Ibarruri ('La Pasionaria'), speech of
    farewell to the International Brigades,
    Barcelona, 1938

5   And I remember Spain
    At Easter ripe as an egg for revolt and ruin.
    Louis MacNeice, 'Autumn Journal' (1939)

6   Deep in the winter plain, two armies
    Dig their machinery, to destroy each other.
    Stephen Spender, 'Two Armies'
    (*c.*1939)

7   The outcome of the Spanish war was settled
    in London, Paris, Rome, Berlin — at any
    rate not in Spain.
    George Orwell, 'Looking Back on the
    Spanish Civil War' (1943)

8   The Spanish Civil War was the Spanish
    share in the tragic European breakdown of
    the twentieth century, in which the liberal
    heritage of the nineteenth century, and the
    sense of optimism which had lasted since
    the Renaissance, were shattered.
    Hugh Thomas, *The Spanish Civil War*
    (1977)

9   The Spanish Republic's struggle was the
    most heroic thing of our lifetime.
    V.G. Kiernan, quoted in MacDougall,
    Voices from the Spanish Civil War (1986)

WAR OF THE SPANISH SUCCESSION
See 167. BLENHEIM, 186. MALPLAQUET

204 BATTLE OF STALINGRAD 1942–3

1   The defeat at Stalingrad threw a scare into
    both the German people and its army. Never
    before in the history of Germany had such a
    large number of troops suffered such a
    terrible fate.
    General Siegfried Westphal, *The German
    Army in the West* (1951)

2   By their incomparable bravery and devotion
    to duty, the officers and men of the army
    raised a memorial to German arms which,
    though not of stone and bronze, will
    nonetheless survive the ages.
    Field Marshal Erich von Manstein, *Lost
    Victories* (1958)

3   In a race with death, which had no trouble
    in catching up with us and was wrenching
    its victims out of our ranks in great batches,
    the army was increasingly pressed into a
    narrow corner of hell.
    Joachim Wieder, *Stalingrad* (1962)

4   The success of the Soviet forces and their
    courageous struggle against the enemy
    inspired all mankind and instilled
    confidence in ultimate victory over Fascism.
    Marshal Georgi Konstantinovich Zhukov,
    quoted in Samsonov, *Stalingrad* (1968)

5   After Stalingrad everything was different.
    After Stalingrad the Russians knew they
    were going to win the war and the Germans
    (except for Hitler) strongly suspected they
    might lose.
    Harrison E. Salisbury, *Marshal Zhukov's
    Greatest Battles* (1969)

205 BATTLE OF THERMOPYLAE 480 BC

1   If the Persians hide the sun, we shall have
    our battle in the shade.
    Dieneces of Trachis, on being told that the
    arrows of the Persian archers would hide the
    sun during the Battle of Thermopylae,
    480 BC, quoted in Herodotus, *Histories*,
    VII, 227, trs. de Selincourt

2   No one could count the number of the dead.
    Herodotus, *Histories*, VII, 220, trs. de
    Selincourt

3 Here lies Megistias, who died
   When the Mede passed Sperchius' tide.
   A prophet; yet he scorned to save
   Himself, but shared the Spartans' grave.
   Megistias, epitaph for himself after the
   Battle of Thermopylae, quoted in
   Herodotus, *Histories*, VII, 227, trs. de
   Selincourt

4 Go tell the Spartans, you who read:
   We took their orders and are dead.
   Simonides of Ceos, epitaph to the Spartan
   dead at the Battle of Thermopylae, quoted
   in Herodotus, *Histories*, VII, 227, trs. de
   Selincourt

5 Earth! render back from out thy breast
   A remnant of our Spartan dead!
   Of the three hundred grant but three,
   To make a new Thermopylae!
   Lord Byron, *Don Juan*, III (1821)

206 THIRTY YEARS WAR 1618–48

1 God is Spanish and fights for our nation
   these days.
   Count Gaspar de Guzman Olivares, letter
   to the Count of Gondomar, 1625

2 In this war His Majesty must have three
   things only in view: namely, to protect his
   kingdoms and hereditary lands so that war
   may be kept from them, then to compel the
   Empire to a just and profitable peace, and
   finally to keep the armies in pay until one
   side or another lays down its arms.
   Albrecht von Wallenstein, letter to the
   Emperor Ferdinand II, 1626

3 All the wars that are on foot in Europe have
   been fused together and have become a
   single war.
   King Gustavus Adolphus of Sweden, 1628,
   quoted in the *Cambridge Modern History*
   (1906)

4 We may indeed say that we have conquered
   our lands from others, and to that end
   ruined our own.
   Count Gustav Oxenstierna, report on the
   progress of the war, 1643, quoted in
   Roberts, *The Swedish Imperial Experience*
   (1979)

5 Something you never believed in
   Has come to pass. What?
   Will the camel pass through the Needle's
        Eye
   Now that peace has returned to Germany?
   Johann Vogel of Nuremberg, verse on the
   Peace of Westphalia, 1648

6 It is the outstanding example in European
   history of meaningless conflict.
   C.V. Wedgwood, *The Thirty Years War*
   (1938)

7 With all its luck and all its danger
   The War is dragging on a bit
   Another hundred years or longer
   The common man won't benefit.
   Bertolt Brecht, *Mother Courage and her
   Children*, 12 (1941)

8 A period in which fervent Christians were
   prepared to hang, burn, torture, shoot or
   poison other fervent Christians with whom
   they happened to disagree upon the correct
   approach to eternal life.
   General Sir John Hackett, *The Profession of
   Arms* (1983)

207 BATTLE OF TOBRUK 1942

1 Corridor to Tobruk clear and secure.
   Tobruk is as relieved as I am.
   General Sir A.R. Goodwin-Austen,
   despatch to Winston S. Churchill, 1941

2 The whole Empire is watching your
   steadfast and spirited defence of this
   important outpost of Egypt with gratitude
   and admiration.
   Winston S. Churchill, message to General
   Morshead in Tobruk, 1941

3 Tobruk! It was a wonderful battle.
   Field Marshal Erwin Rommel, letter to his
   wife, 1942

4 This bloody town's a bloody cuss,
   No bloody trams, no bloody bus,
   And no one cares for bloody us.
   Oh bloody! Bloody! Bloody!
   Hugh Patterson, 'Tobruk', quoted in Laffin,
   Digger (1959)

208 VIETNAM WAR 1954–75

1 We are not going to send American boys
nine or ten thousand miles away from home
to do what Asian boys ought to be doing for
themselves.
Lyndon B. Johnson, campaign speech, 1964

2 Keep asking me no matter how long —
On the war in Vietnam I sing this song —
I ain't got no quarrel with the Viet Cong.
Muhammed Ali, on refusing to be drafted
into the US Army, 1966

3 There's no better way to fight than goin' out
to shoot VCs. An' there's nothing I love
better than killin' Cong. No sir.
General James F. Hollingsworth, interview
with Nicholas Tomalin, in the *Sunday
Times,* 1966

4 I never felt I was fighting for any particular
cause. I fought to stay alive and I killed to
keep from being killed.
David Parks, *GI Diary* (1968)

5 In the fierce fight against the US aggressors,
the most cruel and barbarous imperialist
ringleaders of the present war, our
Vietnamese people will certainly achieve
complete victory, because our great national
salvation resistance has been glowing with
just cause.
General Vo Nguyen Giap, *Big Victory,
Great Task* (1968)

6 Tell me lies about Vietnam.
Adrian Mitchell, 'To Whom it May
Concern', *Out Loud* (1968)

7 The war in Vietnam is mainly unpopular
because it is a war with limited aims, a
police action, and Americans do not identify
with the police.
Frederick A. Pottle, 1968, quoted in
Panichas, *Promise of Greatness* (1968)

8 (Of the Tet Offensive) What the hell is going
on? I thought we were winning the war.
Walter Cronkite, on CBS News, 1968

9 I would hope, myself, that every American
of draft age came across the border here, or
went to Sweden; then they wouldn't have
anyone to fight their wars.
William Kunstler, address in Toronto,
1969, quoted in Williams, *The New Exiles*
(1971)

10 Television brought the brutality of war into
the comfort of the living room. Vietnam was
lost in the living rooms of America — not in
the battlefields of Vietnam.
Marshal McLuhan, interview in the
Montreal Gazette, 1975

11 To win in Vietnam we will have to
exterminate a nation.
Dr Benjamin Spock, *On Vietnam* (1976)

12 The U.S. has broken the second rule of war.
That is, don't go fighting on the mainland of
Asia. Rule One is don't march on Moscow. I
developed these two rules myself.
Field Marshal Viscount Montgomery,
quoted in Chalfont, *Montgomery of
Alamein* (1976)

209 WARS OF THE ROSES 1455–85

1 The slaughter of men was immense: for
besides the dukes, earls, barons and
distinguished warriors who were cruelly
slain, multitudes almost innumerable of the
common people died of their wounds. Such
was the state of the kingdom for nearly ten
years.
Anon., *Historiae Croylandensis Continuatio*
(*c.*1470), trs. Riley in *Ingulph's Chronicel*
(1854)

2 No man was sure of his life, land ne
livelihood, ne of his wife, daughter ne
servant, every good woman standing in
dread to be ravished and defouled.
Act for the Settlement of the Crown upon
Richard III and his Issue, 1484, *Rotuli
Parliamentorum ut et Petitiones et Placita in
Parliamento,* VI (1832)

3 This year, that is to mean ye 18 day of
February, the Duke of Clarence and second
brother to the king, then being prisoner in ye
Tower, was secretely put to death and
drowned in a barrel of malvesye within the
said Tower.
Robert Fabyan, *The Concordance of
Chronicles* (1516)

4 (Of the Battle of Towton, 1461) That battaile weakened wonderfully the force of Englande, seeing those who were killed had been able, both for number and force, to have enterprised in any forreigne war.
Polydore Vergil, *Angliae Historiae Libri* (1534–55)

5 Now is the winter of our discontent
Made glorious summer by this sun of York.
William Shakespeare, *Richard III,* I, i

6 England hath long been mad, and scarred herself;
The brother blindly shed the brother's blood, the father rashly slaughtered his own son, The son, compelled, been butcher to the sire: All this divided York and Lancaster.
William Shakespeare, *Richard III,* V, v

7 Warwick, peace;
Proud setter up and puller down of kings.
William Shakespeare, *Henry VI Part 3,* III, iii

8 The most ferocious and implacable quarrel of which there is factual record.
Winston S. Churchill, *A History of the English Speaking Peoples,* vol. I (1956)

momentous at issue in the strife.
Archibald Alison, *History of Europe* (1842)

5 (Of the British Army) Silently, like the Greeks of old, the men took up their ground, and hardly any sound was heard from the vast array, but the rolling of guns, and an occasional word of command from the officers.
Archibald Alison, *History of Europe* (1842)

6 Waterloo is a battle of the first rank, won by a captain of the second.
Victor Hugo, *Les Misérables* (1862)

7 *Waterloo! Waterloo! Waterloo! morne plaine.*
Victor Hugo, 'L'Expiation'

8 Nine English out of ten think of Waterloo as a purely British victory in which the army of the king of Prussia figures, if it figures at all, as a merely subsidiary factor.
Sir Herbert Maxwell, in *Nineteenth Century* (1900)

YPRES (3RD BATTLE OF)
See 193. PASSCHENDAELE

210 BATTLE OF WATERLOO 1815

1 It has been a damned serious business. Blücher and I have lost 30,000 men. It has been a damned nice thing — the nearest thing you ever saw in your life.
Duke of Wellington, 1815, quoted in Creevey, *Creevey Papers* (1903)

2 It was the most desperate business I ever was in; I never took so much trouble about any Battle; & never was so near being beat.
Duke of Wellington, letter to William Wellesley-Pole, 1815

3 As the sun went down, the darkness swallowed not only an army, but an empire.
Private William Wheeler, *Letters,* ed. Liddell Hart (1951)

4 Never were two armies of such fame, under leaders of such renown, and animated by such heroic feelings, brought into contact in modern Europe, and never were interests so

211 ZULU WAR 1879

1 (Of the Zulus) Still they came on like ants, as a man fell another taking his place, in perfect silence.
Charles Commeline, letter to his parents, 1879, quoted in Emery, *The Red Soldier: Letters from the Zulu War* (1977)

2 I consider that there never was a position where a small force could have made a better defensive stand.
Lt-Gen Lord Chelmsford, memorandum written after the Battle of Isandhlwana, 1879

3 We have certainly been seriously under-rating the power of the Zulu army.
Lt-Gen Lord Chelmsford, after his defeat by the Zulus at the Battle of Isandhlwana, 1879

4 An assegai has been thrust into the belly of
   the nation.
   King Cetewayo, after his victory at the
   Battle of Isandhlwana, 1879

5 To visualise the scene under the lightening
   sky of dawn — the piled Zulu dead, the
   exhausted handful of white men and the
   burned-out buildings behind them — is to
   understand why the name of Rorke's Drift
   will never be forgotten as long as gallant
   deeds are honoured amongst men.
   Wilfred Robertson, *Great Exploits* (1940)

III
ARMIES AND SOLDIERS

AMERICA
See 261. UNITED STATES OF AMERICA

212 ARCHERS

1 Not always will the bow hit that at which it is aimed.
Horace, *De Arte Poetica*, I

2 (Of the Battle of Hattin, 1187) The first Moslem line let fly a cloud of arrows, like a cloud of grasshoppers.
Ibn al-Athir, quoted in *Recueils des Historiens des Croisades* (1872–1906)

3 The Englishmen had their bows y-bent,
Their hearts were good enow;
The first of arrows that they shot off
Seven score spearmen they slew.
Anon., ballad, 'Chevy Chase', 15th century

4 (Of the Battle of Crécy, 1346) Then the English archers stepped forward one pace, and let fly their arrows so regularly, that it appeared like snow.
Jean Froissart, *Chronicles* (1523–5)

5 Wait not while your foe fits arrow to bowstring when you can send your own arrow into him.
Babur, *Babur-nama* (1526–30), trs. Beveridge

6 A bolt lost is not a bow broken.
Sir Walter Scott, *Kenilworth* (1821)

7 In battle, a broken bow has all the makings of a dead archer.
Weston Martyr, 'Bowmen's Battle', *Blackwood's Magazine*, 1938

8 (Of the Battle of Agincourt) The singing of the arrows would not have moved ahead of their flight, but the sound of their impact must have been extraordinarily cacophonous, a weird clanking and banging on the bowed heads and backs of the French men-at-arms.
John Keegan, *The Face of Battle* (1976)

213 ARMIES

1 An army is a beautiful thing, because of the order and well-disposed ranks that are within it.
Richard Stibbes, address to Parliament, 1646, quoted in Walzer, *Revolution of the Saints* (1966)

2 It is not big armies that win battles, it is the good ones.
Marshal Maurice de Saxe, *Mes Rêveries* (1757)

3 An armed, disciplined body is, in its essence, dangerous to liberty. Undisciplined, it is ruinous to society.
Edmund Burke, speech in the House of Commons, 1790

4 The Services in war time are fit only for desperadoes, but in peace are fit only for fools.
Benjamin Disraeli, *Vivian Grey* (1827)

5 A mysterious fraternity born out of smoke
   and danger of death.
   Stephen Crane, *The Red Badge of Courage*
   (1893)

6 The terrible power of a standing army may
   usually be exercised by whomever can
   control its leaders, as a mighty engine is set
   in motion by the cranking of a handle.
   Winston S. Churchill, *The River War* (1899)

7 An army is to a chief what a sword is to a
   soldier.
   Marshal Ferdinand Foch, *Precepts* (1919)

8 The functions of an Army are: 1. to defeat
   the enemy's main force; 2. seize upon his
   vitals.
   General Sir Ian Hamilton, *The Soul and
   Body of an Army* (1921)

9 The army is a crowd — a homogenous
   crowd, it is true, but retaining, despite its
   organisation, some of the general
   characteristics of crowds; intense emotional
   suggestibility, obedience to leaders etc.
   These factors must be handled by
   commanders.
   Gustave Le Bon, *World in Revolt* (1924)

10 The Army, for all its good points, is a
   cramping place for a thinking man.
   Captain Sir Basil Liddell Hart, *Thoughts on
   War* (1944)

11 An army exists to advance by force or the
   threat of force civil policies that cannot be
   advanced by civil methods.
   John Masters, *Bugles and a Tiger* (1956)

12 The deterrence of war is the primary object
   of the armed forces.
   Maxwell D. Taylor, *The Uncertain Trumpet*
   (1960)

13 The principal armed service of this country
   — in its professional attitudes, its officer
   corps — is an extension, a reflection, of that
   country's whole society.
   Correlli Barnett, *The Swordbearers* (1963)

14 Whoever has the army has power, and war
   decides everything.
   Mao Tse-Tung, *Selected Military Writings*
   (1963)

---

214 ARTILLERY

---

1 Artillery is useful to an army when the
   soldiers are animated by the same valour as
   that of the Romans, but without that it is
   perfectly inefficient, especially against
   courageous troops.
   Niccolo Machiavelli, *Discorsi*, XVII (1531)

2 Then let us bring our light artillery,
   Minions, falc'nets, and sakers, to the trench,
   Filling the ditches with the walls' wide
         breach.
   And enter in to seize upon the gold.
   Christopher Marlowe, *Tamburlaine the
   Great*, Part 2, III, iii

3 The cannons have their bowles full of wrath,
   And ready mounted are they to spit forth
   Their iron indignation.
   William Shakespeare, *King John*, II, i

4 Artillerie, th' infernall intrument,
   New-brought from hell, to scourge
         mortalitie
   With hideous roaring and astonishment.
   Samuel Daniel, *The Civil Wars between the
   two Houses of York and Lancaster*, VI
   (1595)

5 *Ultima ratio regum.* The final argument of
   kings.
   Motto inscribed on French cannon by order
   of King Louis XIV, *c*.1660

6 We are the boys
   That fear no noise
   Where the thundering cannons roar.
   Oliver Goldsmith, *She Stoops to Conquer*
   (1773)

7 Then shook the hills with thunder riven,
   Then rushed the steed, to battle driven,
   And louder than the bolts of Heaven,
   Far flash'd the red artillery.
   Thomas Campbell, 'Hohenlinden', *Poems*
   (1803)

8 It is with artillery that war is made.
   Napoleon I, after the Battle of Löbau, 1809

9 My God! I can scarcely credit their taking
   the fearful responsibility of sending us into
   the field practically unarmed with artillery.
   Field Marshal Earl Kitchener, letter to
   Pandeli Ralli during the Boer War, 1900

10 I have seen war, and faced modern artillery,
   and I know what an outrage it is against
   simple men.
   T.M. Kettle, *The Ways of War* (1915)

11 Six evil-looking mouths would spit and
   crack in deafening unceasing clamour, and

the shells pour over their bigger sisters like flocks of birds in whirring flight.
Frankling Lushington, *The Gambardier* (1930)

12 It is of great value to an army, whether in defence or offence, to have at its disposal a mass of heavy batteries.
Winston S. Churchill, *The Gathering Storm* (1948)

13 The remarkable thing about modern shelling is not how many it kills, but how few.
Fred Majdalany, *The Monastery* (1950)

---

215 AUSTRALIA

1 (Of the Australian soldier) The bravest thing God ever made.
Anon. British officer, quoted in *Punch*, 1915

2 (Of Gallipoli) In one day, 25th April, Australia attained nationhood by the heroism of her noble sons.
Phillip F.E. Schuler, *Australia in Arms* (1916)

3 Open-hearted, ever generous, true as gold, and hard as steel, Australia's first great volunteer army, and its valorous deeds, will live in history while the world lasts.
Phillip F.E. Schuler, *Australia in Arms* (1916)

4 The whole Australian army became automatically graded into leaders and followers according to the individual merits of every man, and there grew a wonderful understanding between them.
General Sir John Monash, *The Australian Victories in France* (1920)

5 What these men did nothing can alter now. The good and the bad, the greatness and the smallness of their story will stand. Whatever of glory it contains nothing can now lessen.
C.E.W. Bean, of the 1st Australian Imperial Force, *The Official History of Australia in the War of 1914–1918* (1921)

6 Leadership counts for something, of course, but it cannot succeed without the spirit, elan and morale of those led. Therefore I count myself the most fortunate of men in having

been placed at the head of the finest fighting machine the world has ever known.
General Sir John Monash, of the Australian forces of the First World War, in the *Melbourne Argus*, 1927

7 If we should have to get out, we shall have to fight our way out. No surrender and no retreat.
Maj-Gen Leslie Morshead, Order to his troops (9th Australian Division), at Tobruk, 1941

8 No surrender for me.
Walter Ernest Brown, VC, last words in Malaya, 1942, quoted in Wigmore, *They Died Mightily* (1963)

9 (Of the Battle of Alamein) After the battle I went to see General Morshead, the Australian commander, to congratulate him on the magnificent fighting carried out by his division. His reply was the classic understatement of all time. He said: 'Thank you, General. The boys were interested.'
Lt-Gen Sir Brian Horrocks, *A Full Life* (1960)

10 In the course of the forbidding central period of the Battle of the Somme, Haig's army received a powerful accession of strength and a potent new weapon. The accession of strength was marked by the entry into the battle of the 1st Australian Division opposite Pozieres on July 20th.
John Terraine, *Douglas Haig: The Educated Soldier* (1963)

11 And the band played Waltzing Matilda
as the ship pulled away from the quay
And 'midst all the tears, the flag-waving and cheers
We sailed off to Gallipoli.
Eric Bogle, 'And the Band Played Waltzing Matilda' (1985)

---

216 BLACK AND TANS
See also 236. IRELAND

1 On the 28th day of November
Outside the town of Macroom,
The Tans in their big Crossley tenders
Were hurtling away to their doom.
Anon., song, 'The Boys of Kilmichael'

2 They were a force calculated to inspire

terror in a population less timid and law abiding than the Irish.
Frank O'Connor, *The Big Man* (1937)

3 They consisted of ticket-of-leave men, city toughs, soldiers unable to settle down and ambitious nonentities who had failed to get on — rogues, fools and disappointed men.
C. Desmond Greaves, *Liam Mellowes and the Irish Revolution* (1971)

4 The Black and Tans left an indelible imprint on Anglo–Irish relations.
Robert Kee, *The Green Flag*, vol. III (1972)

BOWMEN
See 212. ARCHERS

217 CANADA

1 I want to show what the Canadian militia can do.
Sir Joseph Carron, Order to the Canadian Militia during Northwest Rebellion, 1885

2 Give me one million men who can hit a target at five hundred yards and we would not have a foe who could invade our country.
Sir Sam Hughes, statement, 1914, quoted in Allen, *Ordeal By Fire: Canada 1910–1945* (1961)

3 The Canadians never budge.
Sir Edwin Alderson, on taking command of the Canadian Corps, 1915, quoted in Beaverbrook, *Canada in Flanders* (1916)

4 It was Canada from the Atlantic to the Pacific on parade. I thought then, and I think today, that in those few minutes I witnessed the birth of a nation.
Brig-Gen Alexander Ross, of the Canadians at the Battle of Vimy Ridge, 1917

5 Who am I? Just another lousy (literally lousy) private in the Canadian Army.
Robert E. Sherwood, quoted in Brown, *The Worlds of Robert E. Sherwood: Mirror to His Time* (1965)

6 I place my trust in the Canadian Corps, knowing that when Canadians are engaged, there can be no giving way.
Sir Arthur Currie, Order of the Day for the final Battle of the Somme, 1918

7 Men of Canada, fight on, carry on; and victory will be yours and ours, for the whole world and for generations yet unborn.
Samuel Gompers, address to the Canadian Club of Ottawa, 1918

8 Above them are being planted the maples of Canada, in the thought that her sons will rest the better in the shade of the trees they knew so well in life.
Arthur Meighen, dedication at Vimy Ridge Memorial, 1921

9 The Canadian Army is a dagger pointed at the heart of Berlin.
General A.G.C. McNaughton, speech in London, 1941

10 Obviously the operation lacked surprise.
Maj-Gen J.H. Roberts, despatch after the failure of the Dieppe Raid, 1942

11 Canada is an unmilitary community. Warlike her people have often been forced to be; military they have never been.
C.P. Stacey, *Official History of the Canadian Army in the Second World War*, vol. I (1955)

12 The real fact is that we prepare for war like precocious giants and for peace like retarded pygmies.
Sir Lester Pearson, acceptance speech, Nobel Prize for Peace, Oslo, 1957

218 CAVALRY

1 The Indians thought the horse and rider were one creature, for they had never seen them before.
Bernal Diaz del Castillo, *The Conquest of New Spain* (1568), trs. Cohen

2 Such a set of ruffians and imbeciles you never beheld, you may call them cannon fodder, but never soldiers. None of them have sat a horse, and when they get their swords I fear they will cut their horses' heads off rather than the enemy's.
Captain Richard Pope, letter to Thomas Coke describing Marlborough's cavalry, 1703

3 Their clothing is of blue, boys, and turned up with red,

A glittering cap and feathers for to adorn
    their head;
A gallant horse to ride boys, with eight
    guineas advance,
If you'll go with Captain Starkey to clip the
    pride of France.
Anon., song of 1798, 'Light Horse'

4 In the face of new weapons, artillery and
infantry have had to alter their tactics, their
modes of action and of fighting; and cavalry
must do the same. Sooner or later cavalry
will retake, among the arms on the field of
battle, the glorious place it has held in the
past.
Colonel Lonsdale Hall, *A Talk to Cavalry
NCOs* (1890)

5 The charge is the culminating point of
cavalry instruction. Rapidity and vehemence
of attack must be united with perfect order
and cohesion.
Field Marshal Earl Haig, *Cavalry Drill*, Part
II (1896)

6 In one respect a cavalry charge is very like
ordinary life. So long as you are all right,
firmly in your saddle, your horse in hand,
and well armed, lots of enemies will give
you a wide berth. But as soon as you have
lost a stirrup, have a rein cut, have dropped
your weapon, are wounded, or your horse is
wounded, then is the moment when from all
quarters enemies rush upon you.
Winston S. Churchill, *The River War* (1899)

7 As well let a blind man out without a dog, as
Infantry without some horsemen to attend
and reconnoitre for it!
Field Marshal Earl Haig, Diary, 1900

8 The enemy are afraid of the British cavalry
and I hope when you get them into the open
you will make an example of them.
Field Marshal Earl Roberts, address to
French's Cavalry Division, Boer War, 1900

9 It must be accepted as a principle that the
rifle, effective as it is, cannot replace the
effect produced by the speed of the horse,
the magnetism of the charge, and the terror
of cold steel.
Field Marshal Earl Haig, *Cavalry Training*
(1907)

10 The real business of cavalry is to manoeuvre
your enemy as to bring him within effective
range of the corps artillery of your own side.
Cecil Chisholm, *Sir John French: An
Authentic Biography* (1915)

11 I disliked the idea of a lot of good horses
being killed and wounded.
Siegfried Sassoon, *Memoirs of an Infantry
Officer* (1930)

12 When one is encumbered with an army
rifle it is no easy matter to clamber on to a
restive sixteen-hand horse in a smart and
soldier-like manner.
J.M. Brereton, 'Stand to Your Horses',
Blackwood's Magazine, 1948

13 The disappearance of the horse from war
does not suggest the withdrawal of riding
from an officer's education in peace. It
suggests exactly the opposite.
General Sir John Hackett, lecture to Royal
United Services Institute, 1960

14 It is hard to be a coward in the midst of a
cavalry charge.
Jenni Calder, *There Must Be a Lone Ranger*
(1974)

---

219 CHINA

1 It seems quite useless to kill the Chinese. It is
like killing flies in July.
Sidney Smith, letter to George Philips, 1842

2 The Red Army fights not merely for the sake
of fighting but in order to conduct
propaganda among the masses, arm them,
and help them establish revolutionary
political power.
Mao Tse-Tung, *Manifestations* (1929)

3 The Chinese soldier is commonly regarded
as a joke by foreigners and as a pest by
compatriots.
Peter Fleming, *One's Company: A Journey
to China* (1934)

4 (Of the Chinese First Front Red Army)
Tactics are important, but we could not
exist if the majority of the people did not
support us. We are nothing but the fist of
the people beating their oppressors!
General P'eng Teh-huai, interview, 1937,
quoted in Snow, *Red Star Over China*
(1938)

5 (Of Chiang Kai-shek's army, 1947) The
whole Chinese army, from top to bottom,
was riven with inhumanity, double-crossing
and terror.
Jack Belden, *China Shakes the World* (1949)

6 Since our Army is the army of the people, it must regularly maintain close relations with the masses, and nourish itself with what is acquired from the struggles waged by the masses. Unless this is done, the Army will be like a tree that has no root, a stream whose source has dried up, and will lose its vitality and combat strength.
General Hsiao Hua, article in *Hung Ch'i,* 1959

7 It is the peasants who are the source of the Chinese army. The soldiers are peasants in military uniform.
Mao Tse-Tung, *Selected Works,* III (1964)

8 (Of the Korean War) Most Americans expected Chinamen to be dwarves. They found themselves assaulted by units which included men six feet and over.
Max Hastings, *The Korean War* (1987)

---

220 COMMANDOS

---

1 Let your plans be dark and impenetrable as the night, and when you move, fall like a thunderbolt.
Sun Tzu, *The Art of War,* VII, 19 (*c.*500 BC)

2 (Of the Boer Commandos) Mounted upon their hardy little ponies, they possessed a mobility which practically doubled their numbers and made it an impossibility ever to outflank them. As marksmen they are supreme.
Arthur Conan Doyle, *The Great Boer War* (1900)

3 Everyone in the army is competing feverishly to get into a commando and it is more glorious to be a subaltern here than a captain in the RM Brigade. It is also a great deal more enjoyable.
Evelyn Waugh, letter to his wife, 1940

4 They had a gaiety and independence which I thought would prove valuable in action. The whole thing was a delightful holiday from the Royal Marines.
Evelyn Waugh, diary after being appointed to 8 Commando, 1940

5 From now on all men operating against German troops in so-called Commando operations are to be annihilated to the last man.
Adolf Hitler, 'Commando Order', issued after the Dieppe Raid, 1942

6 From what I have seen in different parts of the world, forces of this nature tend to be so-called 'Private Armies' because there have been no normal formations to fulfil this function — a role which has been found by all commanders to be a most vital adjunct to their plans.
Brigadier J.M. Calvert, memorandum 'Future of SAS Troops', 1945, quoted in Strawson, *A History of the SAS Regiment* (1984)

7 Any well-trained infantry battalion should be able to do what a commando can do; in the Fourteenth Army they could and did.
Field Marshal Viscount Slim, *Defeat into Victory* (1956)

8 The stars who emerged from commando soldiering were not confined to the officer class or to the famous regiments.
Lord Lovat, *March Past* (1973)

COMMISSIONS
See 247. OFFICERS

---

221 COMMUNICATIONS

---

1 Fighting with a large army under your command is nowise different from fighting with a small one: it is merely a question of instituting signs and signals.
Sun Tzu, *The Art of War,* V, 2 (*c.*500 BC)

2 The important secret of war is to make oneself master of the communications.
Napoleon I, *Maxims of War* (1831)

3 An army must have but one line of operations. This must be maintained with care and abandoned only for major reasons.
Napoleon I, *Maxims of War* (1831)

4 A great part of the information obtained in war is contradictory, a still greater part is false, and by far the greatest part is of doubtful character.
Karl von Clausewitz, *On War* (1832)

5 The line that connects an army with its base of supplies is the heel of Achilles — its most vital and vulnerable point.
John S. Mosby, *War Reminiscences* (1887)

6 Communications dominate war; broadly considered, they are the most important single element in strategy, political or military.
Admiral A.T. Mahan, *The Problem of the East* (1900)

7 The cloud of unknowing which descended on a First World War battlefield at zero hour was accepted as one of its hazards by contemporary generals.
John Keegan, *The Face of Battle* (1976)

222 COSSACKS

1 Let us row, brothers, up the River Volga,
Let us cross, brothers, the steep mountains,
Let us make our way to the Mussulman
            Tsardom,
Let us conquer the Tsardom of Siberia,
And subdue it, brothers, to the White Tsar.
Anon., 'Song of Ermak Timofeevich', c.1580, quoted in Chadwick, *Russian Heroic Poetry* (1932)

2 We serve for grass and water, not for land
            and estates.
Anon., traditional 'Stenka Razin' song

3 The king who rules the Cossacks, rules the world.
Stendhal, *A Life of Napoleon* (1818)

4 These people live as nature does; they die, are born, intermingle, give birth to children, fight, drink, eat, are happy and die again, and their lives are not conditioned except as nature herself inexorably rules the sun, the grass, the animals, the trees and them.
Count Leo Tolstoy, *The Cossacks* (1852–63)

5 We don't need a constitution.
We don't want a republic.
We won't betray Russia.
We will defend the Tsar's throne.
Anon., traditional Cossack song, 1905

6 Their bodies rotted on the fields of Galicia and Eastern Prussia, in the Carpathians and Roumania, wherever the ruddy flames of war flickered and the traces of cossack horses were imprinted in the earth.
Mikhail Sholokhov, *And Quiet Flows the Don* (1929)

DOCTORS
See 243. MEDICAL SERVICES

223 ENGINEERS

1 For 'tis sport to have the engineer
Hoist with his own petar.
William Shakespeare, *Hamlet,* III, iv

2 Good engineers are so scarce, that one must bear with their humours and forgive them because we cannot be without them.
Lord Galway, report from Spain to the Board of Ordnance, 1704, quoted in Chandler, *The Art of Warfare in the Age of Marlborough* (1976)

3 The engineer must be outstandingly bold or outstandingly prudent.
Chevalier de Guignard, *L'Ecole de Mars,* II (1725)

4 Their science demands a great deal of courage and spirit, a solid genius, perpetual study and consummate experience in all the arts of war.
Sébastien de Vauban, *A Manual of Siegecraft* (1740)

5 Sappers often intoxicate themselves at the sap-head, and are then slaughtered like beasts as they take no care what they do; this must be prevented by allowing them only well-watered wine.
Sébastien de Vauban, *Traité de l'Attaque des Places* (1779)

6 'Quo fas et gloria'
Is your motto from today,
And the meaning's: 'Do your duty like a
            Sapper, oh!'
Captain Ward, 'Speech by Sergeant Smith: A Song of the Engineers', quoted in Farmer, *Scarlet and Blue* (1896)

7 I have stated it plain, an' my argument's
            thus
('It's all one,' says the Sapper)

There's only one Corps which is perfect —
    that's us;
An' they call us Her Majesty's Engineers,
Her Majesty's Royal Engineers,
With the rank and pay of a Sapper!
Rudyard Kipling, 'Sappers', *Verse* (1940)

8 Sappers are very elusive as they have no
regiments of their own but get attached to
others like hermit crabs.
Evelyn Waugh, letter to his wife, 1941

9 If there is any truth in the saying that genius
is akin to madness then one might expect a
higher than average incidence of eccentricity
among Sappers. That this has been so is
suggested by the old Army saying that all
Sappers are 'mad, married or Methodist'.
Maj-Gen Frank Richardson, *Mars Without
Venus* (1981)

11 (Of the Royal Engineers) No Corps was
more constantly in demand, so much master
of so many tasks.
General Sir David Fraser, *And We Shall
Shock Them* (1983)

---

224 ENGLAND

---

1 These people, descended from the ancient
Saxons (the fiercest of men) are always by
nature eager for battle, and they could only
be brought down by the greatest valour.
William of Poitiers, quoted in *Gesta
Willielmi Ducis Normannorum et Regis
Anglorum* (*c.*1071), trs. Douglas and
Greenaway

2 This royal throne of kings, this scepter'd
    isle,
This earth of majesty, this seat of Mars,
This other Eden, demi-paradise,
This fortress built by Nature for herself
Against infection and the hand of war.
William Shakespeare, *Richard II*, II, i

3 The blood of the English shall manure the
    ground
William Shakespeare, *Richard II*, IV, I

4 We are Englishmen: that is one good fact.
Oliver Cromwell, speech to Parliament,
1656

5 I have heard much of the English cavalry,

and find it indeed to be the best appointed
and finest I have ever seen.
Prince Eugene of Savoy, quoted in Coxe,
Memoirs of the Duke of Marlborough
(1847)

6 I know these English; they will die on the
ground on which they stand before they lose
it.
Marshal Soult, of the Battle of Waterloo,
quoted in Alison, *The History of Europe*
(1842)

7 *Wellington*: Stand fast, 95th, we must not
be beat: what would they say in England.
95th: Never fear, Sir, we know our duty.
Duke of Wellington, Order to the 95th Foot
during the Battle of Waterloo, 1815

8 We are Englishmen and pride ourselves
upon our deportment, and that pride shall
not be injured in my keeping.
Duke of Wellington, remark after the Battle
of Waterloo, quoted in Mercer, *Journal of
the Waterloo Campaign* (1870)

9 The English soldier is brave, nobody more
so, and the officers generally men of honour.
Napoleon I, quoted in O'Meara, *Napoleon
in Exile*, vol. I (1822)

10 I made a mistake about England in trying to
conquer it. The English are a brave nation.
Napoleon I, quoted in Bryant, *The Age of
Elegance* (1961)

11 Shall not England still be in the van, as she
has always been? Never yet has she failed in
a good cause, and never will she. Has she
not ever struck for Freedom and the Cross?
— inseparable watchwords, that the
experience of the world has taught us must
go hand in hand, or not at all; and where
she strikes, good faith, she drives home well.
George Whyte-Melville, *General Bounce*
(1854)

12 The river of death has brimmed his banks,
And England's far and Honour a name,
But the voice of a schoolboy rallies the
    ranks:
'Play up! play up! and play the game!'
Sir Henry Newbolt, 'Vitaë Lampada' (1897)

13 How can I live among this gentle
obsolescent breed of heroes, and not weep?
Keith Douglas, 'Aristocrats' (1943)

14 The young gentlemen of England do their
    best, of course.

They always do their best, particularly in wartime.
John Mulgan, *Report on Experience* (1947)

15 Even now in the twentieth century, the English prefer a man who goes into the army as a bus driver and comes out as a brigadier. They can still think of him as an amateur, as one of themselves.
Marjorie Ward, *The Blessed Trade* (1971)

16 The English soldier was dogged, enduring, and his courage was often rather unassuming.
Roy Palmer, *The Rambling Soldier* (1977)

225 FRANCE

1 Frenchmen have an unlimited capacity for gallantry and indulge in it on every occasion.
Molière, *The Sicilian* (1666), trs. Wood

2 It is the national characteristic of a Frenchman to attack all the time. His courage rises as he advances towards the enemy, but fades if he is kept waiting; a passive role never suits him.
Lazare Carnot, remark, 1794

3 Every French soldier carries in his cartridge-pouch the baton of a marshal of France.
Napoleon I, quoted in Blaze, *La Vie Militaire sous l'Empire* (1840)

4 The French soldier is indefatigable whenever he pursues a retreating enemy. I repeatedly took advantage of this characteristic.
Napoleon I, quoted in Herold, *The Mind of Napoleon* (1955)

5 A single battle lost takes away a Frenchman's strength and courage, weakens his trust in his superiors and pushes him to insubordination.
Napoleon I, quoted in Herold, *The Mind of Napoleon* (1955)

6 The Army belongs to no party, it belongs to France.
General Galifett, Order of the Day on the conclusion of the Dreyfus Affair, 1898

7 *Au fond*, they are a low lot, and one always has to remember the class these French generals mostly come from.
Field Marshal Sir John French, letter to Earl Kitchener, 1914

8 *Ils ne passeront pas.* They shall not pass.
Anon., watchword during the defence of Verdun, 1916

9 Old France, bowed under the weight of history, bruised by war and revolution, coming and going between grandeur and decline, yet always raised up by the genius of renewal.
General Charles de Gaulle, *Mémoires*, III (1959)

10 Thanks to the Revolution, the French officer corps was mostly middle class and professional, with an oil-and-vinegar admixture of aristocracy.
V.G. Kiernan, quoted in Foot (ed.), *War and Society* (1973)

226 FRENCH FOREIGN LEGION

1 The Legion's in Magenta; the job is in the bag!
Marshal MacMahon, following the Battle of Magenta, 1859

2 They are not men, but devils!
Colonel Francisco Milan, following the defeat of the Mexican army at Camerone, 1863

3 You legionnaires are soldiers in order to die and I am sending you where you can die.
Colonel François de Negrier, Order to French Foreign Legion before embarking for Indo-China, 1883

4 Has not this foreigner become a son of France,
Not by blood inherited, but spilled?
Pascal Bonetti, lines written in 1914

5 The French, a people of etiquette, imagine that the Legion is full of criminals and barefoot savages.
Colonel Ferdinand Maire, letter from his deathbed, 1951, quoted in Geraghty, *March or Die* (1986)

227 GENERALS

1 The principal task of the general is mental, involving large projects and major arrangements.
King Frederick II of Prussia, *Instructions for his Generals* (1747)

2 We should try to make war without leaving anything to chance. In this lies the talent of a general.
Marshal Maurice de Saxe, *Mes Rêveries* (1757)

3 The best generals are those who have served in the artillery.
Napoleon I, letter to General Gaspard Gourgaud, 1817

4 In war the general alone can judge of certain arrangements. It depends on him alone to conquer difficulties by his own superior talents and resolution.
Napoleon I, *Maxims of War* (1831)

5 Generals have never risen from the very learned or erudite class of officers, but have been mostly men who, from the circumstance of their position, could not have attained to any great amount of knowledge.
Karl von Clausewitz, *On War* (1832)

6 The prize of the general is not a bigger tent, but command.
Oliver Wendell Holmes, speech in new York, 1913

7 Battles are lost and won by generals, not by the rank and file.
Marshal Ferdinand Foch, *Precepts* (1919)

8 The perfect general would know everything in heaven and earth.
T.E. Lawrence, letter to Captain Sir Basil Liddell Hart, 1933

9 Efficiency in a general, his soldiers have a right to expect; geniality they are usually right to expect.
Field Marshal Earl Wavell, *Generals and Generalship* (1939)

10 No dead general has ever been criticised, so you have that way out always.
General George S. Patton, letter to his son, 1944

11 I have always believed that a motto for generals must be 'No regrets', no crying over spilt milk.
Field Marshal Viscount Slim, *Defeat into Victory* (1956)

12 Nothing raises morale better than a dead general.
John Masters, *The Road Past Mandalay* (1961)

13 A general is just as good or as bad as the troops under his command make him.
General Douglas MacArthur, address to Congress, 1962

228 GERMANY

1 Christianity has somewhat softened the brutal German lust of battle, but could not destroy it.
Heinrich Heine, *History of Religion and Philosophy in Germany* (1833)

2 Dear Fatherland, may peace be thine!
Dear Fatherland, may peace be thine!
Firm stands, and sure, the watch, the watch on the Rhine!
Firm stands, and sure, the watch, the watch on the Rhine!
Karl Wilhelm, 'Die Wacht am Rhein' (1854), national hymn of Germany during the Franco–Prussian War

3 Prussia, with all the veils that hide a thing, is a military organisation led by a military corporation.
Charles Ardant du Picq, *Battle Studies* (1870)

4 We Germans fear God and nothing else on earth; and it is the fear of God, and nothing else, that makes us love and desire peace.
Count Otto von Bismarck, speech to the Reichstag, 1888

5 Their natural gait is the goose-step.
H.L. Mencken, *Prejudices,* II (1920)

6 Newspaper libels on Fritz's courage and efficiency were resented by all trench-soldiers of experience.
Robert Graves, *Goodbye to All That* (1929)

7 The families of the Rhenish and Westphalian peasants, who made up most of our strength, were glad to let their sons serve in the Army. They were taught habits of punctuality, good behaviour, cleanliness and a sense of duty, which made them better members of the community.
Franz von Papen, *Memoirs,* trs. Connell (1952)

8 (Of the German soldier) They are very brave and tough, and have a marked sense of duty and discipline. Furthermore, they take pride

in mastering their weapons and learning
their job on the battlefield.
Field Marshal Earl Alexander, *Memoirs*
(1962)

9 Whatever we may feel about the Germans,
we must admit that German soldiers were
extremely tough and brave.
Field Marshal Earl Alexander, *Memoirs*
(1962)

10 A German soldier is expected to die rather
than indulge in carelessness with army
property.
Guy Sajer, *Forgotten Soldier* (1971)

11 (Of the German Army in 1945) Like the
warriors of the Teutonic tribes of old, they
were resolved if necessary to die where they
stood, should that be necessary to protect
the uprooted population from the eastern
invader.
John Keegan, *Six Armies in Normandy*
(1982)

12 The military performance of the Wehrmacht
throughout the war was seldom equalled
and only on very rare occasions surpassed.
General Sir David Fraser, *And We Shall
Shock Them* (1983)

---

229 GREECE

---

1 By the Eurymedon these lost youth's glory,
Fighting the best bowmen of the Medes;
They died, on land and on swift-sailing
                    galleys,
Leaving valour's loveliest monument.
Inscription for the Athenian dead at the
Battle of Eurymedon, 469 BC, quoted in von
Gaetringen, *Historische Griechische
Epigramme* (1895)

2 We did not flinch but gave our lives to save
Greece when her fate hung on a razor's
edge.
Simonides, inscription on the Cenotaph at
the Isthmus, *c*.460 BC, quoted in Jay, *The
Greek Anthology* (1973)

3 (Of the Battle of Salamis, 480 BC)
Forth sons of Hellas! Free your land, and
                    free
Your children and your wives, the native
                    seats

Of gods your fathers worshipped and their
                    graves,
This is a bout that hazards all you have.
Aeschylus, *Persians*, II, 353, trs. Cookson

4 The reason why Athens has the greatest
name in all the world is because she has
never given in to adversity, but has spent
more life and labour in warfare than any
other state, thus winning the greatest power
that has ever existed in history.
Pericles, oration to the Athenians, 430 BC,
quoted in Thucydides, *Peloponnesian Wars*,
II, 64, trs. Warner

5 The Greeks are strong, and skilful to their
                    strength,
Fierce to their skill and to their fierceness
Valiant.
William Shakespeare, *Troilus and Cressida*,
I, i

6 When Greeks joined Greeks, then was the
tug of war!
Nathaniel Lee, *The Rival Queens*, IV, ii
(1677)

7 The isles of Greece, the isles of Greece!
Where burning Sappho loved and sung,
Where grew the arts of war and peace,
Where Delos rose and Phoebus sprung!
Lord Byron, *Don Juan*, III (1821)

8 The glory that was Greece.
Edgar Allan Poe, 'To Helen' (1836)

9 Greece stood embattled where eagles fly:
Hellas who showed us how to live
Now taught us how to die.
Hugh Laming, 'Zito Hellas', *Oasis*
(1943)

---

230 GUARDS

---

1 I have seen sieges when the companies of
grenadiers had to be replaced several times.
This is easily explained: grenadiers are
wanted everywhere. If there are four cats to
chase, it is grenadiers who are called for and
usually they are killed without any necessity.
Marshal Maurice de Saxe, *Mes Rêveries*
(1757)

2 Gentlemen of the French Guard, fire first!
   Lord Charles Hay, Order to the French
   Guards at the Battle of Fontenoy, 1745

3 But of all the world's great heroes, there's
             none that can compare,
   With the tow, row, row, row, row, row,
             row, for the British Grenadier.
   Anon., song of the 18th century, 'The
   British Grenadier'

4 Those must be the Guards!
   General Sir John Moore, on seeing the
   British Foot Guards during the retreat to
   Corunna, 1809

5 *La Garde meurt, mais ne se rend pas.* The
   Guards die, but do not surrender.
   Baron de Cambronne, attr. when called on
   to surrender at the Battle of Waterloo, 1815.
   Cambronne denied the saying in 1830.

6 Up Guards and at them again!
   Duke of Wellington, attr. order to his men
   at the Battle of Waterloo, 1815

7 It is better that every man of Her Majesty's
   Guards should lie dead upon the field than
   that they should turn their backs upon the
   enemy.
   Field Marshal Sir Colin Campbell, at the
   Battle of Alma, 1854, quoted in Kinglake,
   The Invasion of the Crimea, III (1877–88)

8 They're changing guard at Buckingham
             Palace —
   Christopher Robin went down with Alice.
   A.A. Milne, 'Buckingham Palace', *When We
   Were Very Young* (1924)

9 Yes, we can safely take the Guards for
   granted.
   Ian Hay, *In the King's Service* (1938)

10 We can sleep tonight. The Guards are in
   front of us.
   Gerald Kersh, *They Die with their Boots
   Clean* (1941)

11 (Of Dunkirk) I can say with some pride as a
   Guardsman that every battalion of the
   Foot-guards arrived back in England with
   their complement of personal weapons
   intact. Nor is it a legend that their trousers
   were pressed!
   Field Marshal Earl Alexander, *Memoirs*
   (1962)

12 (Of Frederick William I of Prussia) The
   Grenadiers were his greatest joy.
   Nancy Mitford, *Frederick the Great* (1970)

---

231 GUERRILLA WAR

---

1 It is just as legitimate to fight an enemy in
   the rear as in the front. The only difference
   is in the danger.
   John S. Mosby, *War Reminiscences* (1887)

2 In civil war, no matter to what extent
   guerrillas are developed, they do not
   produce the same results as when they are
   formed to resist an invasion of foreigners.
   S.I. Gusev, *Lessons of Civil War* (1918)

3 Guerrilla strategy is the only strategy
   possible for an oppressed people.
   Kao Kang, quoted in Mao Tse-Tung, *On
   Guerrilla War* (1937)

4 The ability to run away is the very
   characteristic of the guerrilla.
   Mao Tse-Tung, *Strategic Problems in the
   Anti-Japanese Guerrilla War* (1939)

5 Guerrillas never win wars but their
   adversaries often lose them.
   Charles W. Thayer, *Guerrilla* (1963)

6 Guerrilla war is a kind of war waged by the
   few but dependent on the support of the
   many.
   Captain Sir Basil Liddell Hart, Foreword to
   Mao Tse-Tung, *Guerrilla Warfare* (1961)

7 In guerrilla warfare the struggle no longer
   concerns the place where you are, but the
   places where you are going. Each fighter
   carries his warring country between his toes.
   Franz Fanon, *The Wretched of the Earth*
   (1961)

8 In a war of revolutionary character, guerrilla
   operations are a necessary part. This is
   particularly true in war waged for the
   emancipation of a people who inhabit a vast
   nation.
   Mao Tse-Tung, quoted in Griffith, *Mao
   Tse-Tung on Guerilla Warfare* (1961)

9 The guerrilla wins if he does not lose; the
   conventional army loses if it does not win.
   Henry Kissinger, quoted in *The Times*,
   1969

10 Five essential elements can be picked out,
   without which a resistance movement has
   no chance of success: intelligence, security,
   arms, hope and leadership.
   M.R.D. Foot, *War and Society* (1973)

232 GURKHAS

1 Soldiers of small stature and indomitable spirit.
General Sir Hugh Gough, despatch after the Battle of Sobraon, 1846

2 The Gurkhas were merry little chaps and the only native troops with whom British soldiers were friendly enough for joking and playing tricks.
Frank Richards, *Old-Soldier Sahib* (1936)

3 The Gurkha is a soldier of high battle-skill, a world-famed fighting man and respected in every country where men fought alongside us in the last war.
Lt-Gen Sir Francis Tuker, *While Memory Serves* (1950)

4 The Gurkha keeps faith not only with his fellow men but with great spiritual concepts, and above all, with himself.
John Masters, *Bugles and a Tiger* (1956)

5 To serve with a Gurkha soldier under the British Crown was, and is, a rare privilege which nobody who has shared it can ever forget.
Colonel B.R. Mullalay, *Bugle and Kukri* (1957)

6 The Almighty created in the Gurkhas an ideal infantryman, indeed an ideal Rifleman, brave, tough, patient, adaptable, skilled in field-craft, intensely proud of his military record and unswerving loyalty.
Field Marshal Viscount Slim, *Unofficial History* (1959)

7 It was easy to command such people. It was a privilege to be allowed to do so.
Patrick Davis, *A Child at Arms* (1970)

8 Hearing a British Gurkha officer discuss his profession was something like hearing a priest discuss his vocation.
Byron Farwell, *The Gurkhas* (1984)

HIGHLANDERS
See 257. SCOTLAND

HORSES
See 218. CAVALRY

233 INDIA

1 Bravery is the characteristic of the British army in all quarters of the world; but no other quarter has afforded such striking examples of the existence of this quality in the soldiers as the East Indies.
Duke of Wellington, memorandum on the British Troops in India, 1805

2 The Rajput worships his horse, his sword and the sun, and attends more to the Martial Song of the Bard than to the Litany of the Brahmin.
Lt-Col James Tod, *Annals and Antiquities of Rajastan* (1829)

3 The sepoy is a child in simplicity and biddableness, if you make him understand his orders, if you treat him justly and don't pet him overmuch.
Lord Dalhousie, address to East India Company officers, Meerut, 1857

4 In India there is no dearth of soldiers — of any caste or province; wherever our Government requires one soldier, fifty step forward for service.
N.A. Chick, *Annals of the Indian Rebellion* (1859)

5 If the facilities for washing were as great as those for drink, our Indian Army would be the cleanest body of men in the world.
Florence Nightingale, observations on the sanitary state of the Army in India, Royal Commission, 1863

6 We're marchin' on relief over Injia's coral strand,
Eight 'undred fightin' Englishmen, the Colonel and the Band;
Ho! get away you bullock-man, you've 'eard the bugle blowed,
There's a regiment a-comin' down the Grand Trunk Road.
Rudyard Kipling, 'Route Marchin'', *Barrack-Room Ballads* (1892)

7 Indian soldiers, like soldiers of every nationality, require to be led; and history and experience teach us that eastern races (fortunately for us) however brave and accustomed to war, do not possess the qualities that go to make leaders of men.
Field Marshal Earl Roberts, *Forty-One Years in India,* II (1897)

8 Indians of all classes are of any people I know the easiest led when the leader understands their hearts, and the most difficult to manage when he does not.
General Sir James Willcocks, *The Romance of Soldiering and Sport* (1925)

9 With a sound army, India was safe; with a rotten army she was ruined.
Lt-Gen Sir Francis Tuker, *While Memory Serves* (1950)

10 (Of the Indian Army, 1947) What had taken two hundred years to build was dismembered in three months.
Major Mohammed Ibrahim Quereshi, *The First Punjabis* (1958)

11 Let us remember that for years the Poona colonel was a butt for facetious playwrights and novelists for whom it was easier to produce a stock character stuffed by convention than to create a genuine character inspired by excellence.
Compton Mackenzie, *On Moral Courage* (1962)

12 The Indian Army was like a Praetorian Guard of Empire, set apart from public control, and available always for the protection of the inner state.
James Morris, *Pax Britannica* (1968)

13 (Of the Indian Army) Their morale was based primarily on their collective identity as members of an honourable profession subscribing to a code of courage, fidelity, and devotion to duty, expressed in the Urdu word *izzat,* literally 'honour'.
Dominick Graham and Shelford Bidwell, *Tug of War: The Battle for Italy 1943–1945* (1986)

---

234 INFANTRY

1 Infantry is the nerve of an army.
Francis Bacon, *Essays,* XXIX (1625)

2 My Lord Duke wishes the infantry to step out.
Duke of Marlborough, address to his troops before crossing the River Sensée, 1711

3 I have a very mean opinion of the infantry in general. I know their discipline to be bad and their valour precarious. They are easily put into disorder and hard to recover out of

it; they frequently kill their Officers thro' fear & murder one another in their confusion.
General James Wolfe, letter to his father, 1755

4 Men, you are all marksmen — don't one of you fire until you see the whites of their eyes.
Israel Putnam, command to his men at Bunker Hill, 1775, quoted in Frothingham, *History of the Siege of Boston* (1873)

5 We Riflemen obey orders and do not start difficulties.
Lt-Gen Sir Harry Smith, *Autobiography* (1902)

6 The said arm must be livened by an exalted spirit which will impel it till it comes into contact with the enemy which will enable it to surmount the greatest difficulties and to win at all costs.
Anon., *Spanish Infantry Training Manual,* Toledo, 1908

7 Lines of grey, muttering faces, masked with fear,
They leave their trenches, going over the top,
While time ticks blank and busy on their wrists,
And hope, with furtive eyes and grappling fists,
Flounders in mud. O Jesus, make it stop!
Siegfried Sassoon, 'Attack', *Counter-Attack* (1918)

8 Infantrymen are they who in the frosts and storms of night watch over the sleep of camps, climb under fire the highest crests, fight and die, without their voluntary sacrifice receiving the reward of heroism.
General Francisco Franco, *Diario de Una Bandera* (1922)

9 Look at an infantryman's eyes and you can tell how much war he has seen.
William H. Mauldin, *Up Front* (1944)

10 The rifleman fights without promise of either reward or relief.
General Omar N. Bradley, *A Soldier's Story* (1951)

11 The rifleman trudges into battle knowing that the odds are stacked against his survival.
General Omar N. Bradley, *A Soldier's Story* (1951)

12 To a foot soldier war is almost entirely physical.
Louis Simpson, *With Armed Men* (1972)

13 Infantrymen, however well-trained and well-armed, however resolute, however ready to kill, remain erratic agents of death.
John Keegan, *The Face of Battle* (1976)

14 The infantryman's function was to fight on his feet.
General Sir David Fraser, *And We Shall Shock Them* (1983)

---

235 INTELLIGENCE

1 And Moses sent them to spy out the land of Canaan, and said unto them, Get you up this way southward, and go up into the mountain: And see the land, what it is; and the people that dwelleth therein, whether they be strong or weak, few or many.
Numbers 13:18-19

2 It is essential to know the character of the enemy and of their principal officers — whether they be rash or cautious, enterprising or timid, whether they fight on principle or from chance.
Vegetius, *De Re Militari* (378)

3 Nothing is more worthy of the attention of a good general than the endeavour to penetrate the designs of the enemy.
Niccolo Machiavelli, *Discorsi*, XVIII (1531)

4 Intelligence is the Soul of all Publick business.
Daniel Defoe, letter to Robert Harley, 1704

5 In general it is necessary to pay spies well and not be miserly in that respect. A man who risks being hanged in your service merits being well paid.
King Frederick II of Prussia, *Instructions for his Generals* (1747)

6 One should know one's enemies, their alliances, their resources and nature of their country, in order to plan a campaign.
King Frederick II of Prussia, *Instructions for his Generals* (1747)

7 Have good spies; get to know everything that happens among your enemies; sow dissension in their midst. To crush tyranny every method is fair.
Lazare Carnot, letter to General Pichegru, 1793

8 Great part of the information obtained in war is contradictory, a still greater part is false, and by far the greatest part is of a doubtful character.
Karl von Clausewitz, *On War* (1832)

9 In order to conquer that unknown which follows us until the very point of going into action, there is only one means, which consists in looking out until the last moment, even on the battlefield, for *information*.
Marshal Ferdinand Foch, *Precepts* (1919)

10 The unknown is the governing condition of war.
Marshal Ferdinand Foch, *Precepts* (1919)

11 You can never do too much reconnaissance.
General George S. Patton, *War As I Knew It* (1947)

12 In a battle nothing is ever as good or as bad as the first reports of excited men would have it.
Field Marshal Viscount Slim, *Unofficial History* (1959)

13 Without an efficient Intelligence organisation a commander is largely blind and deaf.
Field Marshal Earl Alexander, *Memoirs* (1962)

14 Nothing helps a fighting force more than correct information. Moreover, it should be in perfect order, and done well by capable personnel.
Ernesto 'Che' Guevara, memorandum, 1963

---

236 IRELAND

1 Worthy are the Irish to dwell in this their land,
A race of men renowned in war, in peace, in faith.
Donatus of Fiesole, 9th century, quoted in *Monumenta Germaniae Historica, Poet. Lat. aevi Carol* (1890), trs. de Paor

2 The land of Ireland is sword-land; let all
men be challenged to show that there is any
inheritance to Fiadh Fail except of conquest
and by dint of battle.
Tadhg Dall O Huiginn, Bardic poem to his
patron (*c*.1550)

3 Well they fought for poor old Ireland, and
        full bitter
was their fate;
O what glorious pride and sorrow fills the
        name of
Ninety-eight!
John K. Casey, 'The Rising of the Moon'
(1861)

4 'Tis Ireland gives England her soldiers, her
generals too.
George Meredith, *Diana of the Crossways*
(1885)

5 We are a Fenian Brotherhood, skilled in the
        arts of war,
And we're going to fight for Ireland, the
        land that we adore.
Anon., 'Song of the Fenian Brotherhood'
(19th century)

6 I'll sing you a song, a soldier's song,
With a cheering rousing chorus
As round our blazing camp-fires we throng
The starry heavens o'er us.
Peader Keaney, 'The Soldier's Song' (1907)

7 For the great Gaels of Ireland
Are the men that God made mad,
For all their wars are merry,
And all their songs are sad.
G.K. Chesterton, *The Ballad of the White
Horse*, II (1911)

8 On the ball, London Irish.
Rifleman Frank Edwards, on kicking a
football towards the German lines at the
Battle of Loos, 1915

9 Then here's to the boys of Kilmichael
Who feared not the might of the foe.
The day that they marched into battle
They laid all the Black and Tans low.
Anon., song of 1920s, 'The Boys of
Kilmichael'

10 We have no shiney gaiters and no Sam
        Browne belts to show
But we're able to defend ourselves no matter
        where we go.
We're out for a Republic and to Hell with
        the Free State,

'No Surrender' is the war cry of the First
        Cork Brigade.
Micéal Barrett, attr., song of 1920s, 'The
First Cork Brigade'

11 The old heart of the earth needed to be
warmed with the red wine of battlefields.
Patrick Pearse, *Political Writings* (1952)

12 (Of the Irish troops in Wellington's army)
Although the Irish were hardy and brave,
they were also ignorant, mad for drink,
violent and without self-discipline.
Correlli Barnett, *Britain and Her Army*
(1970)

237 ISRAEL

1 And God spake unto Israel in the visions of
the night, and said, Jacob, Jacob. And he
said, Here am I. And he said, I am God, the
God of thy father: fear not to go down into
Egypt; for I will there make of thee a great
nation.
Genesis 46:2-3

2 The people arose as one man.
Judges 20:8

3 (David's battle with Goliath) This day will
the Lord deliver thee into mine hand; and I
will smite thee, and take thine head from
thee; and I will give the carcases of the host
of the Philistines unto the fowls of the air,
and to the wild beasts of the earth; that all
the earth might know that there is a God in
Israel.
I Samuel 17:46

4 The beauty of Israel is slain upon the high
places: how are the mighty fallen!
II Samuel 1:19

5 Every time you blow up a British arsenal, or
wreck a British jail, or send a British railroad
train sky-high, or rob a British bank, or let
go with your gun and bombs at British
betrayers and invaders of your homeland the
Jews of America make a little holiday in
their hearts.
Ben Hecht, 'Letter to the Terrorists of
Palestine', advertisement in several New
York newspapers, 1947

6 Israel has emerged stronger than before
   from the test of fire and blood.
   Levi Eshkol, speech in the Knesset, after the
   end of the Six Day War, 1967

7 It was not handed to us on a silver platter:
   you have achieved it after heavy fighting
   soaked in blood and sweat.
   Maj-Gen Yitzhak Rabin, address to the
   Israeli army after the Battle for Jerusalem,
   Six Day War, 1967

8 Return your swords to your scabbards but
   keep them ever-ready, for the time has not
   yet come when you can beat them into
   ploughshares.
   General Moshe Dayan, address to the
   Israeli army after the Battle for Jerusalem,
   Six Day War, 1967

9 One does not have to be very sophisticated
   to come to the conclusion, after the bitter
   experience of twenty years, that the only
   people we can depend on for our security
   are ourselves.
   Golda Meir, interview in the *Sunday Times*,
   1969

10 (Of the Yom Kippur War) Israel has every
   right to draw courage and faith for the
   future from its performance in what the
   Israelis may well remember as their war of
   atonement.
   Maj-Gen Chaim Herzog, *The War of
   Atonement* (1975)

238 ITALY

1 We found gold chains and money in
   abundance by reason the Imperialists had
   lain long there, who, though they gathered
   the whole money of the country, yet they
   had not the wit to transport it away being
   silly, simple Italians and without courage,
   the poorest officers that ever could be
   looked on, and unworthy the name of
   soldiers.
   Captain Sir Robert Munro, *Recollections of
   a Scots Officer of a Worthy Scots Regiment
   with Gustavus Adolphus* (1637)

2 Italy must not only be respected, she must
   make herself feared.
   King Vittorio Emanuele II, proclamation to
   the Italian people, 1866, quoted in Mack
   Smith, *Italy* (1959)

3 We can admire nothing else today but the
   formidable symphonies of bursting shells
   and the crazy sculptures modelled by our
   inspired artillery amongst the enemy hordes!
   Filippo Tommaso Marinetti, reporting
   Italy's invasion of Libya in *L'Intransigent,*
   1911

4 Italy shall be greater by conquest,
   purchasing territory not in shame but at the
   price of blood and glory.
   Gabriele D'Annunzio, speech in Genoa,
   1915, quoted in Mack Smith, *Italy* (1959)

5 Italy
   In this, the uniform
   of your soldier, I rest
   as if
   it were the cradle
   of my father.
   Giuseppe Ungaretti, 'Italy' (1919), trs.
   Silkin and McDuff, *Vita d'un Oumo: Poesie*
   (1947)

6 It is not maligning the Italians to describe
   their achievements, from the purely military
   standpoint, as extraordinarily small.
   General Erich von Falkenhayn, *General
   Headquarters 1914–1916* (1919)

7 Mark my words, it does not matter on
   which side Italy begins the war, for at its
   close she will be found playing her historic
   role as the 'Whore of Europe'.
   Field Marshal Walter von Reichenau, of
   Italy in 1934, quoted in Wheeler-Bennett,
   The Nemesis of Power (1953)

8 Mussolini would probably like it most if the
   German troops would fight in Italian
   uniforms!
   Adolf Hitler, replying to Maj-Gen Hans von
   Funck's assessment of the Italin defeats in
   North Africa, 1940, quoted in Irving, *The
   Trail of the Fox* (1977)

9 Some troops of the Italian army defeated by
   the British in 1940 had their kit-bags all
   ready packed for the occasion.
   Shelford Bidwell, *Modern Warfare* (1973)

239 JAPAN

1 Every day when one's body and mind are at
   peace, one should meditate upon being
   ripped apart by arrows, rifles, spears, and

swords, being carried away by surging waves, being thrown into the midst of a great fire, being struck by lightning, being shaken to death by a great earthquake, falling from a thousand-foot cliff, dying of disease, or committing seppuku at the death of one's master. And every day without fail one should consider himself as dead.
Tsunetomo Yamamato, *Hagakure* (The Book of the Samurai) (*c.*1750)

2 If we look into the causes of Japan's success we find it very largely in the soldierly spirit and self-sacrificing patriotism of the whole of the people.
Lord Baden-Powell, letter in the *Eton College Chronicle,* 1904

3 Secrecy in all things is an ineradicable Japanese trait. As a rule they made no display and tolerated no fuss when dispatching material or sending their troops to war.
Bennet Burleigh, *Empire of the East* (1905)

4 Whether I float as a corpse under the waters, or sink beneath the grasses of the mountainside, I willingly die for the Emperor.
Traditional morning chant of Japanese army recruits, *c.*1932

5 (Of the Battle of Imphal) On this one battle rests the fate of the Empire. All officers and men fight courageously!
Maj-Gen Tanaka, Order of the Day, 1944

6 If one should ask you
What is the heart
of Island Yamato,
It is the mountain cherry blossom
Which exhales its perfume in the morning sun.
Motoori Norinaga, poem in honour of the Japanese kamikaze pilots, 1945

7 The Jap takes as much killing as a conger-eel, and is every bit as slippery.
Sir Bernard Fergusson, *The Wild Green Earth* (1946)

8 The individual soldier remained, as I had always called him, the most formidable fighting insect in history.
Field Marshal Viscount Slim, *Defeat into Victory* (1956)

9 The strength of the Japanese Army lay, not in its higher leadership, which, once its career of success had been checked became

confused, not in its special aptitude for jungle warfare, but in the spirit of the individual Japanese soldier.
Field Marshal Viscount Slim, *Defeat into Victory* (1956)

10 While we admired Japanese courage and tenacity, in the way that men admire the instinctive courage of animals, we also feared them, much more than we feared their bullets. Bullets we understood; but the minds and hearts of the Japanese were beyond our understanding.
Charles Carfrae, *Chindit Column* (1985)

11 They were so different from us, so furtive in their ways, that we still thought of them as some savage wild animal, but whatever their beliefs, their ruthless discipline or their fears, if ever a saga of courage and endurance was written by the common soldier, that of the Japanese Twenty-Eighth Army should be included.
Miles Smeeton, *A Change of Jungles* (1962)

---

240 KNIGHTS HOSPITALLERS

1 We must force them back; but first let us go to church and pray, for time spent in praying is not lost time, and God will fight for us meanwhile.
La Valette, Order to the Knights Hospitallers during the Turkish attack on Malta, 1565, quoted in Surius, *Commentarius brevis rerum in orbe gestarum* (1598)

2 Heroes who make the sea their battlefield — always their breasts uncovered to show the enemy they know no fear.
Henri de Lisdam, *L'esclavage du Brave Chevalier François de Ventimille* (1608)

3 You would have called them Princes, they were so well set up, they marched so arrogantly, with so fine a grace.
Pierre de Brantôme, *Recueil d'aucuns discours* (1665)

4 Only brave and good men feel drawn to the life, for they are only a handful against the Turks, and they have to recite one hundred and fifty paternosters a day, even on the battlefield.
Count George Albert Erbach, of the Knights Hospitallers' wars against the Turks, 1700

5 Lambs at the sound of the church bell, lions at the sound of the trumpet.
Fra Bartolomeo del Pozzo, *Hist. della Sac. Rel. di Malta* (1703)

6 If Grand Masters are mortal, one can say that the Religion of St John is immortal.
G. Aubert de Vertot, *Historia des Chevaliers de St Jean de Jerusalem* (1726)

---

241 KNIGHTS TEMPLARS

1 We propose to you wars which carry with them the reward of glorious martyrdom, wars which assure the title to temporal and eternal glory.
Pope Urban II, statement after the capture of Jerusalem, proposing the formation of a Christian knighthood, 1099, quoted in Bernard of Clairvaux, *De Laude Novae Militiae* (*c.*1128)

2 If you want to save your souls, either throw away the belt of knighthood, or proceed boldly as knights of Christ and go speedily to the defence of the oriental Church.
Bishop Baldric of Dole, statement in Clermont, *c.*1110, quoted in Bernard of Clairvaux, *De Laude Novae Militiae* (*c.*1128)

3 So they are seen to be a strange and bewildering breed, meeker than lambs, fiercer than lions. I do not know whether to call them monks or knights because though both names are correct one lacks a monk's gentleness, the other a knight's pugnacity.
Bernard of Clairvaux, *De Laude Novae Militiae* (*c.*1128)

4 Be not afraid to go to battle, but be ready for the crown of martyrdom.
La Règle de Temple, (*c.*1128), ed. de Curzon

5 They neglected to live but they were prepared to die in the service of Christ.
Ekkehard of Aura, quoted in Vitry, *Historia Orientalis seu Hierosolymitana* (1611)

6 This ideal of the well-born man without possessions was embodied in knight-errantry and templardom, and, hideously corrupted as it always has been, it still dominates sentimentally, if not practically, the military and aristocratic view of life.
William James, *The Varieties of Religious Experience* (1903)

---

242 MARINES

1 Land forces are nothing. Marines are the only species of troops proper for this nation. A powerful fleet and 30,000 Marines will save us from destruction, and nothing else.
Maj-Gen Henry Lloyd, *The History of the Late War in Germany* (1779)

2 How much might be done with a hundred thousand soldiers such as these.
Napoleon I, while inspecting the Royal Marine guard on HMS *Bellerophon*, 1815

3 Tell that to the Marines — the sailors won't believe it.
Sir Walter Scott, *Redgauntlet* (1824)

4 The Marines are properly the garrisons of His Majesty's ships, and under no pretence ought they to be moved from a fair and safe communication with the ships to which they belong.
Duke of Wellington, speech in the House of Lords, 1837

5 A life on the ocean wave,
A home on the rolling deep;
Where the scattered waters rave,
And the winds their revels keep!
Epes Sergeant, 'A Life on the Rolling Wave', the regimental march of the Royal Marines, 1847

6 A ship without Marines is like a garment without buttons.
Admiral David D. Porter, US Navy, letter to Colonel John Harris, US Marine Corps, 1863

7 If the Army and the Navy
Ever gaze on Heaven's scenes,
They will find the streets are guarded
By United States Marines.
Anon., 'The Marines' Hymn', *c.*1880

8 The bended knee is not a tradition of our Corps.
General Alexander Vandergrift, address to the Senate Naval Affairs Committee regarding proposals to abolish the US Marine Corps, 1946

9 I have just returned from visiting the
   Marines at the front, and there is not a finer
   fighting organisation in the world.
   General Douglas MacArthur, after the
   Inchon landings, Korea, 1950

10 Ex-Marines are like ex-Catholics or off-duty
   Feds.
   Michael Herr, *Dispatches* (1977)

243 MEDICAL SERVICES

1 He who wishes to be a surgeon should go to
   war.
   Hippocrates, *Corpus Hippocraticum*

2 Daily practice of the military exercises is
   much more efficacious in preserving the
   health of an army than all the art of
   medicine.
   Vegetius, *De Re Militari* (378)

3 Keep two lancets; a blunt one for the
   soldiers, and a sharp one for the officers:
   this will be making a proper distinction
   between them.
   Francis Grose, *Advice to the Officers of the
   British Army* (1782)

4 The great thing in all military service is
   health.
   Admiral Lord Nelson, letter to Dr Moseley,
   1803

5 You medical people will have more lives to
   answer for in the other world than even we
   generals.
   Napoleon I, quoted in O'Meara, *Napoleon
   in Exile* (1822)

6 Bind up their wounds — but look the other
   way.
   Sir W.S. Gilbert, *Princess Ida*, III (1884)

7 Wise men took refuge in the virtues of cold
   water and kept the surgeons at a safe
   distance.
   Sir John Fortescue, *History of the British
   Army*, I (1899)

8 With quiet tread, with softly smiling faces
   The nurses move like music through the
              room.
   Ivor Gurney, 'Ladies of Charity', *Hospital
   Pictures* (1917)

9 The doctors were working with their sleeves
   up to their shoulders and were red as
   butchers.
   Ernest Hemingway *A Farewell to Arms*
   (1929)

10 To the average professional officer, the
   military doctor is an unwillingly tolerated
   non-combatant who takes sick call, gives
   cathartic pills, makes transportation
   troubles, complicates tactical plans and
   causes the water to smell bad.
   Hans Zinsser, *Rats, Lice and History* (1935)

11 Today we are sending you out to battle.
   Only you are not armed with poison gas and
   bayonets, but with antiseptic, chloroform
   and healing hands. For you march not in the
   path of Alexander and Napoleon but of
   Semmelweiss, Lister and Pasteur. Fight your
   war with courage and honesty, young men,
   for yours is a war against death.
   Budd Schulberg, *The Writer's Radio Theater*
   (1941)

12 No male nursing orderly can nurse like a
   woman, though many think they can.
   Field Marshal Viscount Montgomery,
   Memoirs (1958)

13 Physicians and public health officers, like
   soldiers, are always equipped to fight the
   last war.
   René J. Dubos, *The Dreams of Reason*
   (1961)

244 MERCENARIES

1 Ful ofte tyme he hadde the bord bigonne
   Aboven all nacions in Pruce;
   In Lettow hadde he reysed and in Ruce,
   No Cristen man so ofte of his degree.
   Geoffrey Chaucer, *The Canterbury Tales*,
   Prologue (*c*.1387)

2 They are disunited, ambitious, without
   discipline, faithless, bold amongst friends,
   cowardly amongst enemies, they have no
   fear of God, and keep no faith with men.
   Niccolo Machiavelli, *Art of War* (1560), trs.
   Whitehorne

3 T'were sport for us to hear that all the
   World were in combustion, for then we
   could not want work. 'Tis a blessed trade.
   Edmund Verney, letter to his brother from
   Flanders, 1639

4 I had swallowed, without chewing, in Germanie, a very dangerous maxime, which militarie men there too much follow, which was, that soe we serve our masters honestlie, it is no matter what master we serve.
Sir James Turner, *Memoirs of His Own Life and Times* (c.1686)

5 Let us therefore animate and encourage each other, and show the whole world that a Freeman, contending for liberty on his own account is superior to any slavish mercenary on earth.
General George Washington, General Orders, New York, 1776

6 Wretched is he who is killed in battle not in defence of his native land, but by another's enemies and for another people.
Giacomo Leopardi, *All'Italia* (1818)

7 He, who adopts some other country as his own and makes offer of his sword and his blood, is more than a soldier. He is a hero.
Emile Barrault, 1832, quoted in Dumas (ed.), *Memoirs of Garibaldi* (1929)

8 What cared I for their quarrels or whether the eagle under which I marched had one head or two?
William Makepeace Thackeray, *The Luck of Barry Lyndon* (1844)

9 These, in the day when heaven was falling,
The hour when earth's foundations fled,
Followed their mercenary calling
And took their wages and are dead.
A.E. Housman, 'Epitaph on an Army of Mercenaries' (1914)

10 They were professional murderers and they took
Their blood money and impious risks and died.
Hugh MacDiarmid, 'Another Epitaph on an Army of Mercenaries', *Second Hymn to Lenin* (1935)

11 A mercenary who turns traitor expects the firing squad when he is caught.
Lt-Gen Sir Francis Tuker, *While Memory Serves* (1950)

12 The good fighting man who honestly believes himself to be a pure mercenary in arms, doing it all for the money, may have to guard his convictions as vigilantly as any atheist.
General Sir John Hackett, *The Profession of Arms* (1983)

245 MILITIA

1 Raw in the fields the raw militia swarms,
Mouth without hands; maintained at vast expense,
In peace a charge, in war a weak defence.
John Dryden, *Cymon and Iphegenia* (1699)

2 To place any dependence upon militia is assuredly resting upon a broken staff.
General George Washington, letter to the President of Congress, 1776

3 Every citizen should be a soldier, every soldier a citizen.
Edmond Dubois-Crance, address to the National Assembly, Paris, 1789

4 A well regulated Militia, being necessary to the security of a free State, the right of the people to keep and bear Arms, shall not be infringed.
Constitution of the United States of America, Amendment II, 1791

5 For a people who are free, and who mean to remain so, a well organised and armed militia is their best security.
Thomas Jefferson, memorandum to Congress, 1808

6 Every member of society who is fit for war can be taught, along with his other activities, to master the use of weapons, as much as is needed, not for taking part in parades, but for defending the country.
Friedrich Engels, letter to Karl Marx, 1861

7 We would have an army, but an army of citizens and of soldiers, invincible at home, incapable of waging war abroad; an army without militaristic spirit.
Jules Favre, Bill placed before French government, 1867

8 Wars are not won by heroic militia.
Winston S. Churchill, *The Second World War*, vol. II (1949)

9 Everyone a soldier.
Mao Tse-Tung, political slogan, 1959

10 If the enemy dares attack us, they will be drowned in the great sea of the people's armed forces, and there will be no place to bury them.
Liu Hsien-sheng, article in *Kiangsu Ch'un-chung*, 1959

246 NEW ZEALAND

1 (Of the Maoris) When they attack they work themselves up into a kind of artificial courage which does not allow them time to think much.
Sir Joseph Banks, *Journal* (1769)

2 The New Zealander army John Bull is the most John Bullish.
Anthony Trollope, *Australia and New Zealand* (1873)

3 A Maori was capable of slaughtering wounded and prisoners and perhaps eating them afterwards, but he could also leap down into the fire of both sides to save the life of a fallen foe.
Sir John Fortescue, *A History of the British Army,* XIII (1930)

4 (Of the New Zealand forces at Gallipoli, 1915) The men who fought, fought for the Empire, but also for New Zealand. The Empire belonged to the realm of imagination: New Zealand to the realm of experience.
W.P. Morrell, *The Provincial System in New Zealand* (1932)

5 (Of the North Africa campaign, 1943) It was the New Zealanders who broke the German divisions' heart outside Mersa Matruh.
Alan Moorehead, *African Trilogy* (1944)

6 New Zealand soldiers considered that they had a natural right to enjoy life wherever they were, including the battle zone, or be told the reason.
Dominick Graham and Shelford Bidwell, *Tug of War: The Battle for Italy 1943–1945* (1986)

247 OFFICERS

1 Art thou officer? or art thou base, common and popular?
William Shakespeare, *Henry V,* VI, i

2 I had rather have a plain russet-coated captain that knows what he fights for and loves what he knows, than that which you call 'a gentleman' and is nothing else.
Oliver Cromwell, letter to Sir W. Spring and Maurice Barrow, 1643

3 An ensign's usually a young gentleman who passed through all the classes of his education handsomely enough and was ripe for the university, being designed for a clergyman; but unfortunately happening to be caught abed with one of his mother's chambermaids, the scene was changed and the young spark was doomed to the army.
Edward Ward, *Mars Stript of his Armour* (1709)

4 An officer is much more respected than any other man who has as little money.
Samuel Johnson, quoted in Boswell, *The Life of Samuel Johnson* (1791)

5 As to the way in which some of our ensigns and lieutenants braved danger — the boys just come out from school — it exceeds all belief. They ran as at cricket.
Duke of Wellington, quoted in Rogers, *Recollections* (1859)

6 Lord Wellington does not approve the use of umbrellas during enemy's firing, and will not allow the 'gentlemen's sons' to make themselves ridiculous in the eyes of the enemy.
Lord Arthur Hill, order to the Grenadier Guards, 1813

7 Once a captain always a captain.
Thomas Love Peacock, *Crotchet Castle* (1831)

8 I am the very model of a modern
Major-General,
I've information vegetable, animal and mineral,
I know the kings of England, and I quote the fights historical,
From Marathon to Waterloo, in order categorical.
Sir W.S. Gilbert, *The Pirates of Penzance* (1879)

9 The more helpless the position in which an officer finds his men, the more it is his bounden duty to stay and share their fortune, whether for good or ill.
Field Marshal Viscount Wolseley, quoted in Morris, *The Washing of The Spears* (1968)

10 Concerning brave Captains
Our age hath made known.
Rudyard Kipling, 'Great-Heart' (1919)

11 The idle, so-called governing classes cannot
remain long without war. When there is no
war, they are bored.
Anon., *The Soldier's Welfare: Notes for
Officers* (1941)

12 I had no private means, and without some
£300 a year it was impossible to live as an
officer in a cavalry regiment at home.
Field Marshal Sir William Robertson, *From
Private to Field Marshal* (1921)

13 The care of his men is an officer's duty
which he puts before his own comfort.
Anon., *The Soldier's Welfare: Notes for
Officers* (1941)

14 (Of the British Army officers in North
Africa, 1942)
The noble horse with courage in his eye
clean in the bone, looks up at a shellburst:
away fly the images of the shires
but he puts the pipe back in his mouth.
Keith Douglas, 'Aristocrats' (1943)

15 The badge of rank which an officer wears on
his coat is really a symbol of servitude —
servitude to his men.
General Maxwell D. Taylor, *Army
Information Digest* (1953)

16 The Army, it soon became clear to Mike and
myself at Catterick, was the last surviving
relic of feudalism in English society.
David Lodge, *Ginger, You're Barmy* (1962)

17 The officer-corps on the eve of the Boer
War, 1899, was an expression of late
Victorian upper-class society, rich, snobbish
and corsetted by etiquette.
Correlli Barnett, *Britain and Her Army*
(1970)

ORDNANCE
See 214. ARTILLERY

248 OTHER RANKS

1 As a private soldier, you should consider all
your officers as your natural enemies, with
whom you are in a perpetual state of
warfare.
Francis Grose, *Advice to the Officers of the
British Army* (1782)

2 I think that common soldiers are worse
thought of than other men in the same rank
of life — such as labourers.
James Boswell, *The Life of Samuel Johnson*
(1791)

3 The backbone of the Army is the
non-commissioned man!
Rudyard Kipling, 'The 'Eathen',
Barrack-Room Ballads (1892)

4 Gentlemen rankers out on the spree,
Damned from here to eternity.
Rudyard Kipling, 'Gentlemen-Rankers',
Barrack-Room Ballads (1892)

5 Oh, it's Tommy this, an' Tommy that, an'
'Chuck him out, the brute!'
But it's 'Saviour of 'is country' when the
guns begin to shoot.
Rudyard Kipling, *Barrack-Room Ballads*
(1892)

6 He is no boaster, no cut-throat, but just the
trained, battle-stained, first-class fighting
man who does not funk, who does not
crow, who does not hate, who goes about
his fighting as about any other work,
without fuss and without ill-humour.
Robert Blatchford, *My Life in the Army*
(1910)

7 A *soldier*! — a common *soldier*! — nothing
but a body that makes movements when it
hears a shout!
D.H. Lawrence, *Sons and Lovers* (1913)

8 Soldiers are citizens of death's grey land,
Drawing no dividends from time's
tomorrows.
Siegfried Sassoon, 'Dreamers',
Selected Poems (1925)

9 The ordinary soldier has a surprisingly good
nose for what is true and what false.
Field Marshal Erwin Rommel, quoted in
Liddell Hart, *The Rommel Papers* (1953)

10 Forgive me, Sire, for cheating your intent,
That I, who should command a regiment,
Do amble amiably here, O God,
One of the neat ones in your awkward
squad.
Norman Cameron, 'Forgive Me,
Sire', *Forgive Me, Sire* (1945)

11 We are the little men grown huge with death
    Stolid in squads or grumbling on fatigues,
    We held the honour of the regiment
    And stifled our antipathies.
    Alun Lewis, 'From a Play', *Ha!*
    Ha! Among the Trumpets (1945)

12 Every night, I hear
    The patient, khaki beast grieve in his stall,
    His eyes behind the harsh fingers soft as
                                wool.
    Rayner Heppenstall, 'Instead of a
    Carol', *Poems* (1946)

13 We were privates, and who is more carefree?
    Robert Leckie, *Helmet for my Pillow* (1957)

---

249 PARATROOPS

1 Five thousand balloons, capable of raising
   two men each, could not cost more than five
   ships of the line. And where is the Prince
   who can so afford to cover his country with
   troops for its defense, as that ten thousand
   men descending from the clouds might not
   in many places do an infinite deal of
   mischief before a force could be brought
   together to repel them?
   Benjamin Franklin, *Pennsylvania Gazette,*
   1784

2 They jumped into the air just as sea bathers
   might jump into the sea at a crowded
   seaside resort.
   Lt-Col Giffard Martel, report of the Red
   Army parachute descent, Moscow Military
   District autumn manoeuvres, 1936

3 Be as agile as a greyhound, as tough as
   leather, as hard as Krupp steel: and so you
   shall be the German Warrior incarnate.
   General Kurt Student, 'Ten
   Commandments for Paratroopers' (1938)

4 I felt absolutely no reluctance to jump —
   less than in taking a cold bath. But hitting
   the earth was very shocking.
   Evelyn Waugh, letter to his wife, 1943

5 In airborne forces you expect the
   unexpected the whole time — that's part of
   your make-up
   Maj-Gen R.E. Urquhart, on the failure of
   the Arnhem landings, 1944, quoted in
   Hamilton, *Monty: The Field Marshal
   1944–1976* (1986)

6 (Of the airborne landings in Normandy,
   1944) Having to do it for the first time in
   combat is a chastening experience; it gives a
   man religion.
   Maj-Gen J.M. Gavin, *Airborne Warfare*
   (1947)

7 The flower of German manhood was
   expressed in these valiant highly trained and
   completely devoted Nazi parachute troops.
   To lay down their lives on the altar of
   German Glory and World-power was their
   passionate resolve.
   Winston S. Churchill, *The Second World
   War,* vol. III (1950)

8 (Of the British airborne forces at D-Day,
   1944) These men were the torchbearers of
   liberations.
   Chester Wilmot, *The Struggle for Europe*
   (1952)

9 A parachute is merely a means of delivery
   but not a way of fighting.
   Bernard Fall, *Street without Joy* (1964)

10 (Of the American airborne landings at
   D-Day, 1944) Seventeen men hit the ground
   before their chutes had time to open. They
   made a sound like large ripe pumpkins being
   thrown down to burst against the ground.
   Donald Burgett, *Currahee* (1967)

11 By the nature of its role a parachute division
   confronts the first-line troops of the enemy,
   or very shortly attracts them to its landing
   zones.
   John Keegan, *Six Armies in Normandy*
   (1982)

12 The philosophy of the Parachute Regiment
   is that there is nothing you cannot do. There
   are no limits.
   Major Chris Keeble, during the Falklands
   Campaign, 1982, quoted in Hastings and
   Jenkins, *The Battle for the Falklands* (1983)

---

250 PERSIA

1 If we crush the Athenians and their
   neighbours in the Peloponnese, we shall so
   extend the empire of Persia that its
   boundaries will be God's own sky, so that
   the sun will not look down upon any land
   beyond the boundaries of what is ours.
   King Xerxes, speech to his court before the
   attack on Athens, 480 BC, quoted in
   Herodotus, *Histories,* VII, 6, trs. de
   Selincourt

2 When Persians meet in the streets one can

always tell by their mode of greeting whether or not they are of the same rank; for they so not speak but kiss — their equals upon the mouth, those somewhat superior on the cheeks.
Herodotus, *Histories,* I, 133, trs. de Selincourt

3 The Persians, more than any other nation I know of, honour men who distinguish themselves in war.
Herodotus, *Histories,* VII, 238, trs. de Selincourt

4 Our enemies are Medes and Persians, men who for centuries have lived soft and luxurious lives.
Alexander the Great, address to his army before the Battle of Issus, 333 BC, quoted in Arrian, *The Campaigns of Alexander the Great* II, 7 (*c.*AD 150), trs. de Selincourt

5 Behold a people cometh from the north country, and a great nation shall be raised from the sides of the earth. They shall lay hold on bow and spear; they are cruel, and have no mercy; their voice roareth like the sea; and they ride upon horses, set in array as for war against thee, O daughter of Zion.
Jeremiah, 7:22–3

6 They are most gallant warriors, though crafty rather than courageous, and to be feared only at long range.
Ammianus, *Rerum Gestarum Libri,* XXIII, 6, trs. Rolfe

7 Unhappy Persia — that in former age Hast been the seat of mighty conquerors.
Christopher Marlowe, *Tamburlaine the Great* Part 1, I, i

8 The Persians had a most attractive way of surrendering their cities with dignity. The procedure was so uniform that one suspected a drill laid down in Army Regulations.
Field Marshal Viscount Slim, *Unofficial History* (1959)

PRUSSIA
See 228. GERMANY

251 QUARTERMASTERS

1 Without supplies no army is brave.
King Frederick II of Prussia, *Instructions for his Generals,* II (1747)

2 Nobody ever heard of a quartermaster in history.
Nathanael Greene, letter to George Washington, refusing to be appointed Quartermaster-General of the Continental Army, 1778

3 The standing maxim of your office is to receive whatever is offered you, or you can get hold of, but not to part with anything you can keep.
Francis Grose, *Advice to the Officers of the British Army* (1782)

4 What makes the general's task so difficult is the necessity of feeding so many men and animals. If he allows himself to be guided by the supply officers he will never work and his expedition will fail.
Napoleon I, *Maxims of War* (1831)

5 Bad stations and good liquor and long
      service
Have aged his looks beyond their forty-five;
For eight and twenty years he's been a
      soldier;
And nineteenth months of war have made
      him thrive.
Siegfried Sassoon, 'The Quarter-Master', Diary, 1916

6 The onus of supply rests equally on the giver and the taker.
General George S. Patton, *War As I Knew It* (1947)

7 The general must know how to get his men their rations and every other kind of stores needed for war.
Field Marshal Earl Wavell, *Soldiers and Soldiering* (1953)

RANKERS
See 248. OTHER RANKS

252 RECRUITS

1 An army raised without proper regard to the choice of its recruits was never yet made good by length of time.
Vegetius, *De Re Militari* (378)

2 I don't beat up for common soldiers; no I list only grenadiers — grenadiers, gentlemen. (Sergeant Kite)
George Farquhar, *The Recruiting Officer* (1706)

3 Troops are raised by enlistment with a fixed term, without a fixed term, by compulsion some times, and most frequently by tricky devices.
Marshal Maurice de Saxe, *Mes Rêveries* (1757)

4 Into our town a sergeant came,
With ribbons all so fine
A-flaunting in his cap — alas!
His bow enlisted mine.
Thomas Hood, 'Ballad of Waterloo'

5 The country lad is generally the most humble, obedient, and easily governed soldier; but the city-born recruit is, as a rule, the smartest, tidiest, and most easily trained.
John Menzies, *Reminiscences of an Old Soldier* (1883)

6 Well now, my fine fellows, if you will enlist,
A guinea in gold I will slap in your fist,
And a crown in the bargain to kick up the dust
And drink the Queen's health in the morning.
Anon., Irish ballad of the late 19th century, quoted in Joyce, *Old Irish Folk Music and Songs* (1909)

7 The young recruit is 'aughty — 'e drafs from Gawd knows where;
They bid 'im show 'is stockins an' lay 'is mattress square;
'E calls it bloomin' nonsense — 'e doesn't know, no more —
An' then up comes 'is Company an' kicks 'im round the floor!
Rudyard Kipling, 'The 'Eathen', *Barrack-Room Ballads* (1892)

8 When the 'arf-made recruity goes out to the East
'E acts like a babe an' 'e drinks like a beast,
An' 'e wonders because 'e is frequent deceased
Ere 'e's fit for to serve as a soldier.
Rudyard Kipling, 'The Young British Soldier', *Barrack-Room Ballads* (1892)

9 Take your risk of life and death
Underneath the open sky,
Live clean or go out quick —
Lads, you're wanted. Come and die.
Ewart Alan Mackintosh, 'Recruiting' (1917)

10 Body and spirit I surrendered whole
To harsh Instructors — and received a soul.
Rudyard Kipling, 'The Wonder', *Epitaphs of the War* (1919)

11 When a man puts on a soldier's uniform he leaves behind his former snug and complex life for ever. All that filled his life yesterday becomes like shadows in a dream.
Ilya Ehrenburg, *Red Star*, 1941

12 He's five foot two and he's six foot four,
He fights with missiles and with spears,
He's all of thirty-one and he's only seventeen,
Been a soldier for a thousand years.
Buffy Sainte-Marie, 'The Universal Soldier' (1967)

253 REGIMENTS

1 The regiment is the family. The Colonel, as the father, should have a personal acquaintance with every officer and man, and should instil a feeling of pride and affection for himself, so that his officers and men would naturally look to him for personal advice and instruction.
General William T. Sherman, *Memoirs* (1875)

2 The regiment was like a firework, once ignited; it wheezed and banged with a mighty power.
Stephen Crane, *The Red Badge of Courage* (1895)

3 Keep your hands off the Regiments, you iconoclastic civilians who meddle and muddle in Army matters: you are not soldiers and you do not understand them.
Field Marshal Viscount Wolseley, *The Story of a Soldier's Life* (1903)

4 We never wavered because, in the last resort, we were Gordon Highlanders, we were the Highland Division.
Martin Lindsay, *So Few Got Through* (1946)

5 The peace of mind, happiness and energy of

the soldier come from his feeling himself to be a member of a body solidly united for a single purpose.
Wilfred Trotter, *Instincts of the Herd in Peace and War* (1947)

6 It was from this spirit that no man was alone, neither on the field of battle, which is a lonely place, nor in the chasm of death, nor in the dark places of life.
John Masters, *Bugles and a Tiger* (1956)

7 Loyalty to a fine battalion may take hold of a man and stiffen his resolve.
Lord Moran, *The Anatomy of Courage* (1966)

8 Never forget: the Regiment is the foundation of everything.
Field Marshal Earl Wavell, quoted in Fergusson, *The Trumpet in the Hall* (1970)

9 A good Regiment is a school of manhood, honour and selfless co-operation, and in nine cases out of ten, a soldier is a good soldier to the extent that his Regiment or unit has made him so.
Arthur Bryant, *Jackets of Green* (1972)

10 Britain is littered with monuments to regiments and to battles long forgotten by all but the members of the regiments who fought in them.
Byron Farwell, *Queen Victoria's Little Wars* (1973)

11 Regimental Spirit, for centuries the peculiar pride of the British Army, is for us an indispensable support of high morale and anything which might shake it must be deplored.
Maj-Gen Frank Richardson, quoted in Babington, *For the Sake of Example* (1983)

12 A soldier may be able to accept his own death, the destruction of his section, even the annihilation of his battalion, knowing that his regiment will live on as a mystical unity.
Richard Holmes, *Firing Line* (1985)

254 RESERVES

1 Wars are paid for by the possession of reserves.
Thucydides, *Peloponnesian Wars*, I, 141, trs. Warner

2 A seasonable reinforcement renders the success of a battle certain, because the enemy will always imagine it stronger than it really is, and lose courage accordingly.
Napoleon I, *Maxims of War* (1831)

3 Fatigue the opponent, if possible, with few forces and conserve a decisive mass for the critical moment. Once this decisive mass has been thrown in, it must be used with the greatest audacity.
Karl von Clausewitz, *On War* (1832)

4 He's an absent-minded beggar, but he heard
    his country's call,
And his reg'ment didn't need to send to find
    him!
Rudyard Kipling, 'The Absent-Minded Beggar', *Daily Mail*, 1899

5 I have no more reserves. The only men I have left are the sentries at my gates. I will take them with me to where the line is broken, and the last of the English will die fighting.
Field Marshal Sir John French, message to Marshal Foch during the Battle of Ypres, 1914, quoted in Spears, *Liaison 1914*

6 The reserve is a club, prepared, organised, reserved, carefully maintained with a view to carrying out the one act of battle from which a result is expected — the decisive attack.
Marshal Ferdinand Foch, *Precepts* (1919)

7 It is in the use and witholding of their reserves that the great commanders have generally excelled.
Winston S. Churchill, *Painting as a Pastime* (1932)

8 Reinforcements shambled up past the guns with dragging steps and the expressions of men who knew they were going to certain death.
Aubrey Wade, *The War of the Guns* (1936)

9 I would rather he had given me one more Division.
Field Marshal Erwin Rommel, letter to his wife after receiving the Field Marshal's baton from Hitler for his capture of Tobruk, 1942

10 Replacements are spare parts — supplies. They must be asked for in time by the front line, and the need for them must be anticipated in the rear.
General George S. Patton, *War As I Knew It* (1947)

255 ROME

1 *Ave Caesar, morituri te salutant.* Hail
   Caesar, those who are about to die salute
   you.
   Traditional salute of the gladiators before
   the Roman Emperor

2 There is, I trust, no one so sluggish and dull
   as not to be curious how, and because of
   what qualities in Roman government,
   practically the whole inhabited world in less
   than 53 years fell completely under the
   control of Rome — the like of which, it will
   be found, has never happened before.
   Polybius, *Histories,* XXXIX, 11, trs.
   Chambers

3 *Civis Romanus sum.* I am a citizen of Rome.
   Cicero, *In Verrem,* V, lvii

4 *Aliae nationes servitutem pati possunt;
   populi Romani res est propria libertas.* Other
   nations may be able to endure slavery; but
   liberty is the very birthright of the Roman
   people.
   Cicero, *Philippica,* IV, 7

5 Roman civilisation did not die a natural
   death. It was killed.
   André Piganiol, *L'Empire Chrétien*
   (325–95)

6 The Romans conquered all peoples by their
   discipline. In the measure that it became
   corrupted their success decreased.
   Marshal Maurice de Saxe, *Mes Rêveries*
   (1757)

7 It was an inflexible maxim of Roman
   discipline that a good soldier should dread
   his officers far more than the enemy.
   Edward Gibbon, *The Decline and Fall of the
   Roman Empire,* vol. I, (1776)

8 So sensible were the Romans of the
   imperfections of valour without skill and
   practice that, in their language, the name of
   an Army was borrowed from the word
   which signified exercise.
   Edward Gibbon, *The Decline and Fall of the
   Roman Empire,* vol. I, (1776)

9 Cast thy keys, O Rome, into the deep down
                    falling, even to eternity down
                    falling,
   And weep.
   William Blake, 'A Song of Liberty', *The
   Marriage of Heaven and Hell* (1790–93)

10 While stands the Coliseum, Rome shall
                    stand;
   When falls the Coliseum, Rome shall fall;
   And when Rome falls — the World.
   Lord Byron, *Childe Harold's Pilgrimage,* III
   (1816)

11 Oh Tiber! father Tiber!
   To whom the Romans pray,
   A Roman's life, a Roman's arms,
   Take thou in charge this day!
   Lord Macaulay, *Lays of Ancient Rome,*
   LVI–LIX (1842)

12 The Roman, a politician above all, with
   whom war was only a means, wanted
   perfect means. He took into account human
   weakness, and he discovered the legion.
   Charles Ardant du Pica, *Battle Studies*
   (1870)

13 Legate, I come to you in tears — My Cohort
                    ordered home!
   I've served in Britain forty years. What
                    should I do in Rome?
   Here is my heart, my soul, my mind — the
                    only life I know.
   I cannot leave it all behind. Command me
                    not to go!
   Rudyard Kipling, 'The Roman Centurion's
   Song', *A History of England* (1911)

14 The rain comes pattering out of the sky,
   I'm a Wall soldier, I don't know why.
   W.H. Auden, 'Roman Wall Blues',
   Collected Poems (1976)

256 RUSSIA

1 There is only one defence against the
   Russians and that is a very hot climate.
   Stendahl, *A Life of Napoleon* (1818)

2 Russia has two generals in whom she can
   confide — Generals Janvier and Février.
   Tsar Nicholas II of Russia, quoted in
   Punch, 1853

3 The defence of the fatherland is the sacred
   duty of every citizen of the USSR.
   Constitution of the USSR (1936)

4 We Bolsheviks have come from the bowels
   of the people and we prize and love the
   glorious deeds in the history of our people.
   Vyacheslav M. Molotov, speech to the
   Supreme Soviet, 1939

5 The Russian winter was a surprise for the Prussian tourists.
Ilya Ehrenburg, *Red Star,* 1941

6 The German Army in fighting Russia is like an elephant attacking a host of ants. The elephant will kill thousands, perhaps even millions, of ants, but in the end their numbers will overcome him, and he will be eaten to the bone.
Colonel Bernd von Kleist, 1941, quoted in Wheeler-Bennett, *Nemesis of Power* (1953)

7 Comrades! Our forces are numberless. The over-weening enemy will soon learn this to his cost. Side by side with the Red Army many thousand workers, collective farmers and intellectuals are rising to fight the enemy aggressor. The masses of the people will rise up in their millions.
Josef Stalin, address to the people of the Soviet Union, 1941

8 Of these people you can make everything — nails, tanks, poetry, victory.
Editorial in *Soviet War News,* 1942

9 (Of the First World War) The valour and marvellous endurance of her soldiery, badly led and worse armed, saved, though they could not decide, the Allied cause.
B.H. Sumner, *Survey of Russian History* (1944)

10 Only he who saw the endless expanse of Russian snow during this winter of our misery, and felt the icy wind that blew across it, burying every object in its path; who drove for hour after hour through that no man's land only at last to find too thin shelter, with insufficiently clothed half-starved men; and who also saw by contrast the well-fed warmly clad and fresh Siberians, fully equipped for winter fighting; only a man who knew all that can truly judge the events which now occurred.
Colonel-General Heinz Guderian, *Memories of a Soldier* (1951)

11 A Russian infantry attack is an awe-inspiring spectacle; the long grey waves come pounding on, uttering fierce cries, and the defending troops require nerves of steel.
Field Marshal F.W. von Mellenthin, *Panzer Battles* (1955)

12 The stoicism of the majority of Russian soldiers and their mental sluggishness makes them quite insensible to losses. The Russian soldier values his own life no more than those of his comrades.
Field Marshal F.W. von Mellenthin, *Panzer Battles* (1955)

13 Defence, retreat and winter — on these resources the Russian high command relied.
Winston S. Churchill, *A History of the English Speaking Peoples,* vol. III (1957)

14 Russians, in the knowledge of inexhaustible supplies of manpower, are accustomed to accepting gigantic fatalities with comparative calm.
Barbara Tuchman, *The Guns of August* (1962)

15 The Soviet soldier, who had travelled the hard road to the approaches of Berlin, was consumed with hatred for the enemy and wanted only to finish him off as quickly as possible and rid mankind of the threat of fascist enslavement.
Marshal Georgi Konstantinovich Zhukov, 'The Battle of Berlin', *Voyenno-Istoricheski Zhurnal* (1965)

SAPPERS
See 223. ENGINEERS

257 SCOTLAND

1 The Scots are a bold, hardy people, very experienced in war. At that time they had little love or respect for the English and the same is true today.
Jean Froissart, *Chronicles* (1523–5)

2 Scots, wha hae wi' Wallace bled,
Scots wham Bruce has aften led,
Welcome to your gory bed.
Or to victorie.
Robert Burns, 'Robert Bruce's Address to his Army, before the Battle of Bannockburn'

3 See how they wane, the proud files of the Windermere,
Howard — Ah! woe to thy hopes of the day!
Hear the wide welkin rend,
While the Scots' shouts ascend,
'Elliot of Lariston, Elliot for aye!'
James Hogg, 'Lock the Door, Lariston', *The Spy* (1811)

4 Scotland for ever!
   Traditional battle-cry, attr. the Royal Scots
   Greys, before their charge on the French
   lines at Waterloo

5 The Scot will not fight until he sees his own
   blood.
   Sir Walter Scott, *The Fortunes of Nigel*
   (1822)

6 (Of the Scottish soldier) He was taught to
   consider courage as the most honourable
   virtue, cowardice the most disgraceful
   failing; to venerate and obey his chief, and
   devote himself for his native country and
   clan; and thus prepared to be a soldier, he
   was ready to follow wherever honour and
   duty called him.
   General Sir David Stewart of Garth,
   Sketches etc. (1822)

7 It was the writing, quite as much as the
   fighting of the Scottish regiments that
   distinguished them.
   Alexander Somerville, *Autobiography of a
   Working Man* (1848)

8 I am now upwards of sixty years old; I have
   been forty years in the service; I have been
   engaged in actions seven and twenty times,
   but in the whole of my career I have never
   seen any regiment behave so well as the
   78th Highlanders. I am proud of you. I am
   not a Highlander, but I wish I was one.
   Brig-Gen Sir Henry Havelock, address to
   the Seaforth Highlanders after the Relief of
   Lucknow, 1857

9 A trusty sword to draw, Willie
   A comely weird to dree [fate to fulfil],
   For the royal rose that's like the snaw,
   And the King across the sea!
   Andrew Lang, 'Kenmure, 1715'

10 The General says this hill must be taken at
   all costs — the Gordon Highlanders will
   take it.
   Lt-Col Henry Mathias, Order to the
   Gordon Highlanders before the Battle of
   Dargai, 1897

11 Scotland is poorer in men but richer in
   heroes.
   Inscription to the Highland regiments which
   fought at the Battle of Magersfontein, South
   Africa, 1899

12 The Gooks will never drive the Argylls off
   this hill.
   Major Kenneth Muir, VC, last words,

during the Argyll and Sutherland
Highlanders' battle for Point 282, Korean
War, 1950

13 His native stubbornness added to his
   regular discipline and training has made the
   Highlander as steadfast in defence as any;
   but he has always kept the fierceness and
   swiftness in attack for which his ancestors
   were so famous.
   Field Marshal Earl Wavell, *Soldiers and
   Soldiering* (1953)

14 Highland soldiers brought glory to Scotland
   under Wolfe at Quebec, and ever since have
   stood in the forefront of the British Army.
   Winston S. Churchill, *A History of the
   English Speaking Peoples*, vol. III (1957)

15 War is part of the spiritual climate of
   Scottish life.
   John Keegan, *Six Armies in Normandy*
   (1982)

16 The Scottish infantry, splendid fighters as
   they were, were given to riot and
   indiscipline, and took a pride in the fact that
   only their own officers and NCOs could
   control them.
   Dominick Graham and Shelford Bidwell,
   Tug of War: The Battle for Italy 1943–1945
   (1986)

SEPOYS
See 233. INDIA

SOVIET UNION
See 256. RUSSIA

---

258 SPARTA

---

1 (Of the Spartan army) Fighting singly, they
   are as good as any, but fighting together
   they are the best soldiers in the world.
   Demaratus of Sparta, to King Xerxes before
   the invasion of Greece, 480 BC, quoted in
   Herodotus, *Histories*, VII, 107, trs. de
   Selincourt

2 I assure you that if you can defeat these men
   and the rest of the Spartans who are still at

home, there is no other people in the world who will dare to stand firm or lift a hand against you.
Demartus of Sparta, to King Xerxes before the Battle of Thermopylae, 480 BC, quoted in Herodotus, *Histories*, VII, 209, trs. de Selincourt

3 (Of Leonidas and the three hundred at the Battle of Thermopylae, 480 BC) They resisted to the last, with their swords if they had them, and, if not, with their hands and teeth.
Herodotus, *Histories*, VII, 227, trs. de Selincourt

4 Because of our well-ordered life we are both brave in war and wise in council.
King Archidamus, speech to the Spartans before the Peloponnesian War, 432 BC, quoted in Thucydides, *Peloponnesian Wars*, Il, 84, trs. Warner

5 You Spartans are the only people in Hellas who wait calmly on events, relying for your defence not on action but on making people think you will act.
Thucydides, *Peloponnesian Wars*, I, 69, trs. Warner

6 Nearly the whole Spartan army, except for a small part, consists of officers serving under officers.
Thucydides, *Peloponnesian Wars*, V, 66, trs Warner

7 The Spartans on the sea-wet rock sat down
  and combed their hair.
A.E. Housman, 'The Oracle', *Collected Poems* (1939)

SPECIAL FORCES
See 220. COMMANDOS, 249. PARATROOPS

259 STAFF OFFICERS

1 I do not want to make an appointment on my staff except as such as are early risers.
General Thomas Jackson, letter to his wife, 1862

2 The British soldier can stand up to anything except the British War Office.
George Bernard Shaw, *The Devil's Disciple*, III (1901)

3 I am decidedly of the opinion that we cannot have a first-rate army, unless we have a first-rate staff, well educated, constantly practised at manoeuvres, and with wide experience.
Field Marshal Earl Roberts, 1902, quoted in the *Royal Commission on the War in South Africa*

4 If I were fierce, and bald, and short of breath,
I'd live with scarlet Majors at the Base,
And speed glum heroes up the line to death.
Siegfried Sassoon, 'Base Details', *Counter-Attack* (1918)

5 If the relations between the General and his Chief of Staff are what they ought to be, the boundaries are easily adjusted by soldierly and personal tact and the qualities of mind on both sides.
Field Marshal Paul von Hindenburg, *Out of my Life* (1920)

6 The ideal General Staff should, in peace time, do nothing! They deal in an intangible stuff called thought. Their main business consists in thinking out what an enemy may do and what their Commanding Generals ought to do, and the less they clash their spurs the better.
General Sir Ian Hamilton, *The Soul and Body of an Army* (1921)

7 The staff knew so much more of the war than I did that they refused to learn from me of the strange conditions in which Arab irregulars had to act; and I could not be bothered to set up a kindergarten of the imagination for their benefit.
T.E. Lawrence, *The Seven Pillars of Wisdom* (1926)

8 The military staff must be adequately composed: it must contain the best brains in the fields of land, air and sea warfare, propaganda war, technology, economics, politics and also those who know the peoples' life.
General Erich von Ludendorff, *Total War* (1935)

9 My war experience led me to believe that the staff must be the servants of the troops, and that a good staff officer must serve his commander and the troops but himself be anonymous.
Field Marshal Viscount Montgomery, *Memoirs* (1958)

10 There is far too much paper in circulation in the Army, and no one can read half of it intelligently.
Field Marshal Viscount Montgomery, *Memoirs* (1958)

260 TRANSPORT

1 Build no more fortresses, build railways.
Helmuth, Graf von Moltke, maxim adopted in the 1860s

2 Victory is the beautiful bright-coloured flower. Transport is the stem without which it could never have blossomed.
Winston S. Churchill, *The River War* (1899)

3 Without munitions of war my armies cannot fight; without food they cannot live. You are helping to send these things to them each day, and in so doing you are hurling your spears at the enemy.
King George V, address to the South African Native Labour Corps, 1917, quoted in James, *Mutiny* (1987)

4 I sing of the Gordons,
Lament to brave soldiers,
They will not come home to their land and
    their wives,
O Lowlands and Highlands
And all the small islands —
Don't wait for the transport that never
    arrives.
David Martin, 'Lament for the Gordons', after the fall of Singapore, 1942

5 When Hitler put his war on wheels he ran it straight down our alley. When he hitched his chariot to an internal combustion engine, he opened up a new battle front — a front that we know well — it's called Detroit.
Maj-Gen Brehon B. Somervell, address to US Army Service Force, 1942, quoted in Keegan and Holmes, *Soldiers* (1985)

6 Mobility is the true test of a supply system.
Captain Sir Basil Liddell Hart, *Thoughts on War* (1944)

7 The soldier cannot be a fighter and a pack animal at one and the same time, any more than a field piece can be a gun and a supply vehicle combined.
Maj-Gen J.F.C. Fuller, letter to S.L.A. Marshall, 1948

8 The more I see of war, the more I realise how it all depends on administration and transport.
Field Marshal Earl Wavell, *Soldiers and Soldiering* (1953)

9 (Of troopships) Living between decks is as noisy, as crowded, and as uncomfortable as living in a fairground.
Walter Robson, *Letters from a Soldier* (1960)

10 Without the automobile engine, Verdun could not have been saved.
A.J.P. Taylor, *The First World War* (1963)

261 UNITED STATES OF AMERICA

1 'Tis the star-spangled banner, O! long may
    it wave
O'er the land of the free and the home of the
    brave!
Francis Scott Key, 'The Star-Spangled Banner' (1814)

2 If destruction be our lot we must ourselves be its author and founder. As a nation of free men we must live through all time or die by suicide.
Abraham Lincoln, speech at Springfield, Illinois, 1858

3 The men and officers of your command have written the name of Virginia today as high as it has ever been written before.
General Robert E. Lee, letter to Maj-Gen George Pickett, after the failure of Pickett's Charge during the Battle of Gettysburg, 1863

4 There is such a thing as a man being too proud to fight.
President Woodrow T. Wilson, speech in Philadelphia, 1915

5 Let us set ourselves a standard so high that it will be a glory to live up to it, and then let us live up to it and add a new laurel to the crown of America.
President Woodrow T. Wilson, address to the soldiers of the National Army, 1917

6 The man who has not raised himself to be a soldier, and the woman who has not raised

her boy to be a soldier for the right, neither of them is entitled to the citizenship of the Republic.
Theodore Roosevelt, speech to troops at Camp Upton, 1917

7 (Of the First World War) We're here to be killed. How do you want to use us?
General Tasker Howard Bliss, to Marshal Foch, quoted in Aston, *Biography of Foch* (1929)

8 There can be no fifty-fifty Americanism in this country. There is room for only one hundred per cent Americanism.
Theodore Roosevelt, speech at Saratoga, New York, 1918

9 Here rests in honored glory an American soldier known only to God.
Inscription on the Tomb of the Unknown Soldier, Arlington National Cemetery, Washington DC

10 Done give myself to Uncle Sam,
Now I ain't worth a good goddam,
I don't want no mo' camp,
Lawd, I want to go home.
Anon., American Negro song, quoted in Odum, *Wings on My Feet* (1929)

11 We must be the great arsenal of democracy.
President Franklin D. Roosevelt, radio broadcast, 1940

12 (Of the Burma campaign, 1944) The closer you got to the front, the more you found that the Americans, whether or not they saluted with great punctilio, meant business, and the closer became the ties between them and the rest of us.
John Masters, *The Road Past Mandalay* (1961)

13 The American obsession with a sudden attack on the United States by the Soviet Union is an intelligible obsession after Pearl Harbour; but so long as that obsession endures it will be a threat to the peace of the world.
Compton Mackenzie, *On Moral Courage* (1962)

14 Americans think of themselves collectively as a huge rescue squad on twenty-four hour call to any spot on the globe where dispute and conflict may erupt.
Eldridge Cleaver, 'Rallying Round the Flag', *Soul on Ice* (1968)

15 The Army does not exist to serve itself, but it exists to serve the American people.
General E.C. Meyer, speech at Washington, 1981

16 American commanders were overtly romantic about the capability of their men and covertly disgusted at their actual achievements.
Dominick Graham and Shelford Bidwell, *Tug of War: The Battle for Italy 1943–1945* (1986)

262 VOLUNTEERS

1 Every man thinks meanly of himself for not having been a soldier.
Samuel Johnson, quoted in Boswell, *The Life of Samuel Johnson* (1791)

2 Who is the happy warrior? Who is he
That every man in arms should wish to be?
William Wordsworth, 'Character of the Happy Warrior' (1807)

3 Listen, young heroes! your country is
       calling!
Time strikes the hour for the brave and the
       true!
Now while the foremost are fighting and
       falling,
Fill up the ranks that have opened for you!
Oliver Wendell Holmes, *The Last Leaf* (1831)

4 The patriot volunteer, fighting for his country and his rights, makes the most reliable soldier on earth.
General Thomas Jackson, quoted in Henderson, *Stonewall Jackson and the American Civil War* (1898)

5 I'm going to leave my mother, I am going far
       from her;
I know that she will miss me for I am her
       darling boy.
And if ever I return again I'll let them all see
       me;
With a star and medal on my breast a soldier
       brave I'll be.
John J. Blockley, 'The Scarlet and The Blue', popular song of the 1870s

6 A fool you say! Maybe you're right.
I'll 'ave no peace unless I fight.

I've ceased to think; I only know
I've gotta go, Bill, gotta go.
Robert Service, 'The Volunteer', *Rhymes of a Red Cross Man* (1916)

7 Why did we join the Army, boys?
Why did we join the Army?
Why did we come to France to fight?
We must have been bloody well barmy!
Anon., song of the First World War, 1914–18

8 If you are only ready to go when you are fetched, where is the merit of that? Are you only going to do your duty when the law says you must? Does the call to duty find no response in you until reinforced, let us rather say superseded, by the call of compulsion?
Field Marshal Earl Kitchener, speech at the Guildhall, London, 1915

9 Our voluntary service regulars are the last descendants of those rulers of the ancient world, the Roman legionaries.
General Sir Ian Hamilton, *Gallipoli Diary*, I (1920)

10 It was commonly believed that any young man who joined the army did so because he was too lazy to work, or else he had got a girl in the family way.
Frank Richards, *Old-Soldier Sahib* (1936)

11 The volunteer principle is all very well, and those who volunteer deserve honour; but it skims the cream off units.
Brigadier Sir Bernard Fergusson, *The Trumpet in the Hall* (1970)

12 We can build a strong, capable, effective Army with volunteers as long as we have that foundation of pride. We cannot buy pride; we cannot legislate it; we cannot force it, we cannot do without it.
General Creighton W. Abrams, address to National Aviation Club, Washington, 1974

---

263 WALES

---

1 There was many a corpse beside Argoed Llwyfain;
From warriors ravens grew red
And with their leader a host attacked.
For a whole year I shall sing to their triumph.
Taliesin, *Gwaith Argoed Llwyfain* (c.590), trs. Conran

2 Men went to Catraeth, embattled, with a cry,
A hoste of horsemen in brown armour, carrying shields,
Spear-shafts held aloft with sharp points,
And shining mail-shirts and swords.
Aneirin, *The Gododdin* (c.600), trs. Jackson

3 This people is light and active, hardy rather than strong, and entirely bred up to the use of arms; for not only the nobles, but all the people are trained to war.
Giraldus Cambrensis, *Descriptio Cambriae*, I (1193)

4 Though it appear a little out of fashion,
There is much care and valour in this Welshman.
William Shakespeare, *Henry V*, IV, i

5 An old and haughty nation, proud in arms.
John Milton, *Comus* (1634)

6 The chapels held soldiering to be sinful, and in Merioneth the chapels had the last word.
Robert Graves, *Goodbye to All That* (1929)

7 My fathers were with the Black Prince of Wales.
David Jones, *In Parenthesis* (1939)

8 The fabled Dragon banner flies once more above the Dee
Where the sons of Wales are gathering to set our people free
From wrong and dire oppression, pray, my son, for strength anew,
For widows will be weeping at the falling of the dew.
A.G. Prys-Jones, 'A Ballad of Glyndwr's Rising', *Poems of Wales* (1923)

9 To their fate over a far ridge.
These went
Like the bright early youths
That used to go as innocent sacrifices
To the ancient gods of the earth.
Gwilym R. Jones, 'On Dyffryn Nantlle School War Memorial', *Cerddi* (1969)

10 We were a people wasting ourselves
In fruitless battles for our masters,
In lands to which we had no claim,
With men for whom we felt no hatred.
R.S. Thomas, 'Welsh History'

264 WOMEN SOLDIERS

1 (Of the Amazons) They have a marriage law which forbids a girl to marry until she has killed an enemy in battle; some of their women, unable to fulfil this condition, grow old and die in spinsterhood.
Herodotus, *Histories*, III, 119, trs. de Selincourt

2 Neither earth nor ocean
Produces a creature as savage and
             monstrous
As woman.
Euripides, *Hecuba*, trs. Arrowsmith

3 A whole troops of foreigners would not be able to withstand a single Gaul if he called his wife to his assistance, who is usually very strong, and with blue eyes; especially when swelling her neck, gnashing her teeth, and brandishing her sallow arms of enormous size, she begins to strike blows mingled with kicks, as if they were so many missiles sent from the string of a catapult.
Ammianus, *Rerum Gestarum Libri*, XV, 7 (*c.*392), trs. Yonge

4 The wemen war of litil les vassalage and strenth than was the men; for al rank madinnis and wiffis, gif they war nocht with child, yeid als weill to battall as the men.
Hector Boece, *Chronicles of Scotland*, I, 20 (1526), trs. Bellenden

5 No captaine of England; behold in your
             sight
Two brests in my bosome, and therfor no
             knight:
Noe knight, sirs, of England, nor captaine
             you see,
But a poor simple lass, called Mary Ambree.
Anon., ballad, 'Mary Ambree'

6 She listed volunteer in a regiment of foot,
By beating the drum great honour she got;
Twice as much more does she undertake,
For she beats on the drum for her true
             lover's sake.
Anon., song of *c.*1703, 'The Drum Major'

7 As a soldier bold she did enlist with her
             lover by her side,
And no one e'er suspected a red jacket
             concealed a maid;
She fought beside her own true love and
             never did give o'er,
Till they beat their daring enemy, the Sikhs
             on India's shore.
Anon., song of *c.*1850, 'The Indian War'

8 The Germans thought Russian women would do their washing for them in the morning and dance for them in the evening. Those gross insolent males, accustomed to dealing with greedy submissive females, expected to find housewives and fan-dancers. Instead they've found women ready to defend their honour and freedom with the last drop of their blood.
Ilya Ehrenburg, *Red Star*, 1941

9 A woman's place should be in bed and not the battlefield, in crinoline or Terylene rather than in battledress, wheeling a pram rather than driving a tank. Furthermore, it should be the natural function of women to stop men from fighting rather than aiding and abetting them in pursuing it.
John Laffin, *Women in Battle* (1967)

265 YEOMEN OF THE GUARD

1 Wherefore for the saue garde & preseruacion of his owne body he constituted and ordaynd a certain nombre as well of good archers as of diuerse other persons being hardy, strong and of agilitie to giue daily attendance on his person, whome he named Yomen of his garde.
Edward Hall, *The Union of the Noble and Illustre Famelies of Lancastre and York* (1542)

2 These were they that in times past made all France afraid, and albeit they be not called master, as gentlemen are; or Sir as to Knights appertaineth but onlie 'John and Thomas', etc, yet they have been found to have done verie good worke.
Raphael Holinshed, *The Historie of England* (1587)

3 And now whether King Henry doubted any sudden attempt upon his person or whether he did it to follow the example of France; in the very beginning of his Reign he ordained a band of tall personable Men to be attending upon him which was called the King's Guard; which no King before and all Kings since have usd.
Sir Richard Baker, *Chronicles of the Kings of England* (1643)

4 Tower Warders,
Under orders,
Gallant Pikemen, valiant sworders!
Brave in bearing
Foemen staring,
In their bygone days of daring!
Sir W.S. Gilbert, 'Chorus of the Yeomen of the Guard', *The Yeomen of the Guard*, I (1888)

IV
WAR AND PEACE

266 AGGRESSION

1 In peace, there's nothing so becomes a man
As modest stillness and humility;
But when the blast of war blows in our ears,
Then imitate the action of the tiger.
William Shakespeare, *Henry V*, III, i

2 Anger is one of the sinews of the soul.
Thomas Fuller, *The Holy State and the Profane State* (1642)

3 The urge for destruction is also a creative urge!
Michael Bakunin, *Jahrbuch für Wissenschaft und Kunst* (1842)

4 It is not merely cruelty that leads men to war, it is excitement.
Henry Ward Beecher, *Proverbs from the Plymouth Pulpit* (1887)

5 It is better to be violent, if there is violence in our hearts, than to put on the cloak of non-violence to cover impotence.
M.K. Gandhi, *Non-violence in Peace and War* (1948)

6 People who are vigorous and brutal often find war enjoyable, provided that it is a victorious war and that there is not too much interference with rape and plunder. This is a right help in persuading people that wars are righteous.
Bertrand Russell, *Unpopular Essays* (1950)

7 Aggression unchallenged is aggression unleashed.
Lyndon B. Johnson, statement to the nation after North Vietnamese attacks on US warships in the Gulf of Tonkin, 1964

8 An army which thinks *only* in defensive terms is doomed. It yields initiative and advantage in time and space to an enemy — even an enemy inferior in numbers. It loses the sense of the hunter, the opportunist.
General Sir David Fraser, *And We Shall Shock Them* (1983)

267 AIR POWER

1 Aviation is war not a sport.
General de Castelnau, remark in 1918, quoted in Baring, *Flying Corps Headquarters 1914–1918* (1920)

2 In order to assure an adequate national defence, it is necessary — and sufficient — to be in a position in case of war to conquer the command of the air.
Giulio Doulet, *The Command of the Air* (1924)

3 It is probable that future war will be conducted by a special class, the air force, as it was by the armoured knights of the Middle Ages.
William Mitchell, *Winged Defence* (1924)

4 Aeroplanes are most effective against morale. They frighten; they exhaust; they break nerves. They do not, usually, in fact, kill many men.
Tom Wintringham, *English Captain* (1939)

5 The illusion that afflicts all men, however brave, when subjected to their first dive-bombing attack — the terrible ineradicable belief that you yourself have been singled out specially for destruction,

that the diving plane has seen you personally, and is coming straight for you, and that nothing on God's earth can stop it.
Richard Austin ('Gun Buster'), *Return via Dunkirk* (1940)

6 The large ground organisation of a modern air force is its Achilles heel.
Captain Sir Basil Liddell Hart, *Thoughts on War* (1943)

7 Air power is the most difficult of all forms of military force to measure or even to express in precise terms.
Winston S. Churchill, *The Second World War*, vol. I (1948)

8 The function of the Army and Navy in any future war will be to support the dominant air arm.
General James H. Dolittle, speech to the Georgetown University Alumni Association, 1949

9 Modern air power has made the battlefield irrelevant.
Sir John Slessor, *Strategy for the West* (1954)

10 Air attack causes a disproportionate amount of alarm.
Richard Holmes, *Firing Line* (1985)

268 AMBUSH

1 Attack him when he is unprepared, appear when you are not expected.
Sun Tzu, *The Art of War*, I, 24 (*c.*490 BC)

2 An ambuscade, if discovered and promptly surrounded, will repay the intended mischief with interest.
Vegetius, *De Re Militari* (378)

3 Everything which the enemy least expects will succeed the best.
King Frederick II of Prussia, *Instructions for his Generals* (1747)

4 To be defeated is pardonable; to be surprised — never!
Napoleon I, *Maxims of War* (1831)

5 'All quiet along the Potomac,' they said,
'Except now and then a stray picket
Is shot as he walks on his beat to and fro
By a rifleman hid in the thicket.'
Ethel L. Beers, 'The Picket Guard' (1861)

6 A snider squibbed in the jungle —
Somebody laughed and fled.
And the men of the First Shikaris
Picked up their Subaltern dead,
With a big blue hole in his forehead,
And the back blown out of his head.
Rudyard Kipling, 'The Grave of the Hundred Dead', *Barrack-Room Ballads* (1892)

7 They found, who fell to the sniper's aim,
A field of death on the field of fame.
Patrick McGill, 'In the Morning', after the Battle of Loos, 1915, *Soldier Songs* (1917)

8 Inaction leads to surprise, and surprise to defeat, which is after all, only a form of surprise.
Marshal Ferdinand Foch, *Precepts* (1919)

9 War is the realm of the unexpected.
Captain Sir Basil Liddell Hart, *Defence of the West* (1950)

269 AMMUNITION

1 The fatal balls of murdering basilisks.
William Shakespeare, *Henry V*, I, ii

2 Put your trust in God, my boys, and keep your powder dry.
Oliver Cromwell, Order to his troops before the Battle of Marston Moor, 1644

3 I heard the bullets whistle; and believe me, there is something charming in the sound.
General George Washington, letter to his mother after the Battle of Great Meadows, 1754

4 Every bullet hath its billet.
King William III, quoted in Wesley, *Journal* (1765)

5 Better pointed bullets than pointed speeches.
Count Otto von Bismarck, speech during the Hesse-Cassel insurrection, 1850

6 Men died in heaps upon the Aubers Ridge ten days ago, because the field guns were short, and gravely short, of high explosive shells.
Lord Northcliffe, editorial in *The Times* after the Battle of Aubers Ridge, 1915

7　If any mourn us in the workshop, say
We died because the shift kept holiday.
Rudyard Kipling, 'Batteries out of
Ammunition', *Epitaphs of the War* (1919)

8　The most precious thing when in contact
with the enemy is ammunition. He who fires
uselessly, merely to assure himself, is a man
of straw.
General Kurt Student, 'Ten
Commandments for Paratroopers' (1938)

9　The flying bullet down the Pass
That whistles clear, 'All flesh is grass.'
Rudyard Kipling, 'Arithmetic on the
Frontier', *Verse* (1940)

10　Praise the Lord and pass the ammunition.
Chaplain H.M. Forgy, attr., during the
attack on Pearl Harbor, 1941

11　There is nothing more democratic than a
bullet or a splinter of steel.
Wendell L. Willkie, *An American Program*
(1944)

12　Human beings, like plans, prove fallible in
the presence of those ingredients that are
missing in manoeuvres — danger, death and
live ammunition.
Barbara Tuchman, *The Guns of August*
(1962)

270　AMPHIBIOUS OPERATIONS

1　The fleet and the army acting in concert
seem to be the natural bulwark of these
Kingdoms.
Thomas More Molyneux, *Conjunct
Expeditions* (1759)

2　The question of landing in the face of an
enemy is the most complicated and difficult
of the war.
General Sir Ian Hamilton, *Gallipoli Diary*
(1920)

3　It is a crime to have amphibious power and
leave it unused.
Winston S. Churchill, memorandum to the
Chiefs of Staff Committee, 1940

4　In landing operations, retreat is impossible.
General George S. Patton, General Order to
the US 7th Army before the Sicily landings,
1943

5　If anyone had told us two years ago that we
could throw ashore a million men, two
hundred thousand vehicles, and
three-quarters of a million tons of stores,
across open beaches, in none too favourable
weather, in thirty days, we would have
dubbed him mad.
General Lord Ismay, letter to Admiral Earl
Mountbatten after D-Day, 1944

6　If the Battle of Waterloo was won on the
playing fields of Eton, the Japanese bases in
the Pacific were captured on the beaches of
the Caribbean.
Holland M. Smith, *Coral and Brass* (1949)

7　The amphibious landing is the most
powerful tool we have.
General Douglas MacArthur, Planning
Conference for the Battle of Inchon, 1950

8　We went in again at dawn; it was like
motoring through a regatta, and the silence
was like a blanket of lead.
James Cameron, on the Inchon Landings
during the Korean War, *Picture Post*, 1950

9　Surprise, violence, and speed are the essence
of all amphibious landings.
Winston S. Churchill, *The Second World
War*, vol. V (1952)

10　Amphibious flexibility is the greatest
strategic asset that a sea power possesses.
Captain Sir Basil Liddell Hart, *Deterrence
or Defence* (1960)

271　BARRACKS

1　A hut, like a little cottage, for Soldiers to lie
in.
English Military Dictionary (1702)

2　The people of this Kingdom have been
taught to associate the idea of barracks and
slavery so closely together that, like
darkness and the devil, though there be no
connection between them, yet they cannot
separate them, nor think of the one without
thinking at the same time of the other.
Field Marshal George Wade, statement to
the Board of Ordnance, 1740, quoted in
Clode, *The Military Forces of the Crown*, II
(1869)

3　(Of the barracks in Portsmouth) Every kind
of corruption, immorality and looseness is

carried to excess; it is a sink of the lowest and most abominable of vices.
General James Wolfe, letter to Lord George Sackville, 1758

4 You can smell some soldiers' feet before you enter their rooms.
Patrick Jervis, *From Ensign to Ship's Captain* (1812)

5 This den of living abominations.
J. MacMullen, *Camp and Barrack-Room* (1846)

6 The men could only wait and wait, and watch the shadow of the barracks creeping across the blinding white dust.
Rudyard Kipling, *Soldiers Three* (1890)

7 Single men in Barricks don't grow into plaster saints.
Rudyard Kipling, 'Tommy', *Barrack-Room Ballads* (1892)

8 The barrack-square, washed clean with rain, Shines wet and wintry-grey and cold.
Siegfried Sassoon, 'In Barracks', *Counter-Attack* (1918)

9 O dark, musty platoon huts, with the iron bedsteads, the chequered bedding, the lockers and the stools! Even you can become the object of desire; out here you have a faint resemblance to home; your rooms, full of the smell of stale food, sleep, smoke, and clothes.
Erich Maria Remarque, *All Quiet on the Western Front* (1929)

10 Our truck ground and whined through a seemingly endless expanse of squat huts huddled together round bleak parade grounds, forbidding barrack blocks, dejected rows of married quarters, and everywhere obtrusive military notice boards, with their strident colours and barbaric language of abbreviations.
David Lodge, *Ginger, You're Barmy* (1962)

272 BATTLES

1 A day of batell is a day of harvest for the devill.
William Hock, *New England's Tears* (1640)

2 I do not favour battles, particularly at the beginning of a war. I am sure a good general

can make war all his life and not be compelled to fight one.
Marshal Maurice de Saxe, *Mes Rêveries* (1757)

3 The first blow is half the battle.
Oliver Goldsmith, *She Stoops to Conquer* (1773)

4 Battles decide everything.
Karl von Clausewitz, *Principles of War* (1812)

5 Battles, even in these ages, are transacted by mechanism; men now even die, and kill one another, in an artificial manner.
Thomas Carlyle, *The French Revolution* (1837)

6 I felt great anxiety, no fear; curiosity also. It was unpleasant until the fire opened, and then only one idea possessed me: that of keeping the soldiers steady and animated.
General Sir Charles Napier, quoted in Napier, *The Life and Opinions of Sir Charles James Napier GCB* (1857)

7 Noise, tumult, smoke, the sight of blood and agony — these are not enough. Hatred and anger are needed, the spirit of revenge.
Ernest Dawson, 'A Reconnaissance', *Blackwood's Magazine*, 1900

8 The thundering line of battle stands, And in the air Death moans and sings; But Day shall clasp him with strong hands, And Night shall fold him in soft wings.
Julian Grenfell, 'Into Battle', *The Times* (1915)

9 I never enjoyed anything so much in my life — flames, smoke, SOSs, lights, drumming of guns, swishing of bullets all appeared stage properties to set off a majestic scene.
Hugh Quigley, *Passchendaele and the Somme* (1928)

10 We have become wild beasts. We do not fight, we defend ourselves against annihilation.
Erich Maria Remarque, *All Quiet on the Western Front* (1929)

11 The acid test of battle brings out the pure metal.
General George S. Patton, *War As I Knew It* (1947)

12 The battlefield is the epitome of war.
S.L.A. Marshal, *Men Against Fire* (1947)

13 In battle the prudent become daring, the

miserly lavish, and even cowards display valour.
Jean Dutourd, *Taxis of the Marne* (1957)

14 All men wait for battle and when it comes
Pass along the sword's edge their resilient
                thumbs.
Emyr Humphreys, 'An Apple Tree and a
Pig', *Ancestor Worship* (1970)

15 In battle, life reduces itself to primal
simplicity.
Edmond Ions, *A Call to Arms* (1972)

273 BAYONETS

1 The onset of Bayonets in the hands of the
Valiant is irresistable.
Maj-Gen John Burgoyne, Orderly Book,
1777

2 It is an excellent instrument for digging
potatoes, onions or turnips.
Francis Grose, *Advice to the Officers of the
British Army* (1782)

3 When bayonets deliberate, power escapes
from the hands of the government.
Napoleon I, *Political Aphorisms* (1848)

4 Under Divine blessing, we must rely on the
bayonet when firearms cannot be furnished.
General Thomas Jackson, letter requesting
the supply of 1000 pikes, 1862

5 He who loves the bristle of bayonets only
sees in their glitter what beforehand he feels
in his heart.
Ralph Waldo Emerson, *Miscellanies* (1884)

6 This bloody steel
Has killed a man.
I heard him squeal
As on I ran.
Wilfred Gibson, 'The Bayonet', *Battle*
(1916)

7 Sweet Sister, grant your soldier this:
That in good fury he may feel
The body where he sets his heel
Quail from your downward darting kiss.
Siegfried Sassoon, 'The Kiss',
Counter-Attack (1918)

8 The laurels of victory are at the point of the
enemy bayonets. They must be plucked
there; they must be carried by a

hand-to-hand fight if one really means to
conquer.
Marshal Ferdinand Foch, *Precepts* (1919)

9 Man, it seemed, had been created to jab the
life out of Germans.
Siegfried Sassoon, *Memoirs of an Infantry
Officer* (1930)

10 So long as a bare couple of yards separates
the men, the bullet can outreach the
bayonet.
Captain Sir Basil Liddell Hart, *Thoughts on
War* (1944)

11 It is the threat of the bayonet and the sight
of the point that usually does the work. The
man almost invariably surrenders *before* the
point is stuck into him.
Fred Majdalany, *The Monastery* (1950)

12 For most of the Great War it was simply an
anachronism, useful as a toasting fork,
biscuit slicer or intimidator of prisoners.
Denis Winter, *Death's Men* (1978)

13 Bayonet-fighting is indescribable — a man's
emotions race at feverish speed and
afterwards words are incapable of
describing feelings.
I.L. Idriess, quoted in Holmes, *Firing Line*
(1985)

274 BOMBARDMENT

1 No one without experiencing it can imagine
the delight a man feels when, after three
hours' bombardment, he leaves so
dangerous a spot as the lodgements.
Count Leo Tolstoy, *Tales of Army Life*
(1855)

2 (Of the Battle of Passchendaele) It was an
awful pandemonium, something like the
medieval idea of hell, pitch-dark apart from
the evil flashes of bursting shells, screams,
groans and sobs, men writhing in the mud,
men trying to walk and falling down again
and everywhere figures scurrying like lost
souls, backwards and forwards,
blaspheming and imploring someone to tell
them the way.
Major Desmond Allhausen, Diary, 1917,
quoted in Warner, *Passchendaele* (1987)

3 Those whispering guns — O Christ, I want
                to go out

And screech at them to stop — I'm going
    crazy;
I'm going stark, staring mad because of the
    guns.
Siegfried Sassoon, 'Repression of War
Experience' *Counter-Attack* (1918)

4 They say music's the most evocative art in
the world, but *sacré nom de Dieu*, they
hadn't counted the orchestra of a
bombardment.
Ernest Raymond, *Tell England* (1922)

5 Now began an indescribable carnage.
Grenades flew through the air like snowballs
till the whole scene was veiled in white
smoke.
Ernst Junger, *The Storm of Steel* (1929)

6 The shells screamed overhead so thick and
fast they seemed to eclipse the sky as with
an invisible roof, rumbling like earthquakes
behind, crashing like a thousand cymbals
before us, a pillar of cloud against the
dawning east — leading us on!
Charles Carrington, *A Subaltern's War*
(1929)

7 I think it is well for the man in the street to
realise that there is no power on earth that
can prevent him from being bombed.
Stanley Baldwin, speech in House of
Commons, 1931

8 The very earth, lacerated and torn, rocked
and dissolved under the weight and force of
the metal blown into it. Both the earth and
the heavens seemed rent by this
concentrated effort of man's fury.
E.J. Rule, *Jacka's Mob* (1933)

9 There were men in France who were ready
to go out but who could not meet death in
that shape. They were prepared for it if it
came cleanly and swiftly, but that
shattering, crudely bloody end by a big shell
was too much for them.
Lord Moran, *Anatomy of Courage* (1945)

10 With a shell whistling at you there is not
much time to pretend and a person's
qualities are starkly revealed.
Marguerite Higgins, *The War in Korea*
(1951)

11 Winged metal death
Roaring its way
Beneath summer skies.
Harri Gwynn, 'London 1944',
Barddoniaeth (1955)

12 Being shelled is the real work of an infantry
soldier.
Louis Simpson, *With Armed Men* (1972)

13 The agony of the men being shelled began
well before the explosion. The skilled ear
picked out each gun, noted its calibre, the
path of its shell and the likely explosion
point.
Denis Winter, *Death's Men* (1978)

14 Shelling was a severe test for isolated men,
particularly at night when shells made a
louder and more frightening noise and
sounded closer than in fact they were.
Dominick Graham and Shelford Bidwell,
Tug of War: The Battle for Italy 1943–1945
(1986)

15 Few experiences are more calculated to
discourage infantry waiting to attack than a
brisk shelling from their own guns falling
short.
Philip Warner, *Passchendaele* (1987)

---

275 BRAVERY
See also 282. COURAGE

---

1 Even the bravest cannot fight beyond his
strength.
Homer, *The Iliad,* XIII

2 *Fortes fortuna adiuvat.* Fortune favours the
brave.
Terence, *Phormio*

3 Valour is the soldier's adornment.
Livy, *Histories,* IX

4 Few men are brave by nature, but good
order and experience may make many so.
Niccolo Machiavelli, *The Art of War* (1520)

5 Perfect valour consists in doing without
witnesses what one would be capable of
doing before the world at large.
Duc de La Rochefoucauld, *Maxims* (1678),
trs. Pocock

6 None but the brave deserves the fair.
John Dryden, *Alexander's Feast* (1697)

7 Brave actions never want a trumpet.
Thomas Fuller, *Gnomologia* (1732)

8 Nothing in life is so stupid as a gallant
officer.
Duke of Wellington, after the Battle of
Fuentes de Onoro, 1811

9 Whither depart the souls of the brave that
    die in battle,
Die in the lost, lost fight, for the cause that
    perishes with them?
Arthur Hugh Clough, *Amours de Voyage*
(1858)

10 Bravery never goes out of fashion.
William Makepeace Thackeray, *The Four
Georges* (1860)

11 Wars may cease, but the need for heroism
shall not depart from this earth, while man
remains man and evil exists to be redressed.
Rear-Admiral A.T. Mahan, *Life of Nelson*
(1897)

12 I do not believe that there is any man who
would not rather be called brave than have
any other virtue attributed to him.
Field Marshal Viscount Slim, *Courage and
Other Broadcasts* (1957)

13 Ever since the First World War there has
been an inclination to denigrate the heroic
aspect of man.
Compton Mackenzie, *On Moral Courage*
(1962)

---

276 CHIVALRY

1 A Knyght ther was, and that a worthy man,
That fro the tyme that he first bigan
To riden out, he loved chivalrie,
Trouthe and honour, fredom and curteisie.
Geoffrey Chaucer, *The Canterbury Tales*,
Prologue (*c.*1387)

2 The age of chivalry has gone and that of
sophisters, economists and calculators has
succeeded.
Edmund Burke, *Reflections on the French
Revolution* (1790)

3 So faithful in love, and so dauntless in war,
There never was knight like the young
    Lochinvar.
Sir Walter Scott, *Marmion*, V, xii (1808)

4 Forward each gentleman and knight!
Let gentle blood show generous might,
And chivalry redeed the fight!
Sir Walter Scott, *The Lord of the Isles*, VI
(1815)

5 The world's male chivalry has petered out,
But women are knight-errants to the last.
Elizabeth Barrett Browning, *Aurora Leigh*,
IV (1857)

6 And indeed He seems to me
Scarce other than my king's ideal knight,
Who reverenced his conscience as his king;
Whose glory was, redressing human wrong;
Who spake no slander, no, nor listened
    to it;
Who loved one only and who clave to her.
Lord Tennyson, *Idylls of the King*
(1859–85)

7 My good blade carves the casques of men,
My tough lance thrusteth sure,
My strength is as the strength of ten,
Because my heart is pure.
Lord Tennyson, 'Sir Galahad', *Idylls of the
King* (1859–85)

8 Chivalry is an ingredient
Sadly lacking in our land.
Sir, I am your most obedient,
Most obedient to command!
Sir W.S. Gilbert, *The Sorcerer*, I (1877)

9 Some say that the age of chivalry is past,
that the spirit of romance is dead. The age of
chivalry is never past, so long as there is a
wrong left unredressed on earth.
Fanny Kingsley, *Life of Charles Kingsley*
(1879)

10 In one sense the charge that I did not fight
fair is true. I fought for success and not for
display. There was no man in the
confederate army who had less of the spirit
of knight-errantry in him, or who took a
more practical view of war than I did.
John S. Mosby, *War Reminiscences*
(1887)

11 Honour has come back, as a king, to earth,
And paid his subjects with a royal wage;
And Nobleness walks in our ways again;
And we have come into our heritage.
Rupert Brooke, 'The Dead', *1914 and
Other Poems* (1915)

12 It is the essence of chivalry to interfere.
Philip Guedalla, *Fathers of the Revolution*
(1926)

13 For killing to be gentlemanly, it must take
place between gentlemen.
John Keegan, *The Face of Battle* (1976)

277 COLOURS

1 Fair as the moon, clear as the sun, and
terrible as an army with banners.
Song of Solomon 6:10

2 The soldiers should make it an article of
faith never to abandon their standard.
Marshal Maurice de Saxe, *Mes Rêveries*
(1757)

3 Stood for his country's glory fast,
And nail'd her colours to the mast.
Sir Walter Scott, *Marmion*, I (1808)

4 Every means should be taken to attach the
soldier to his colours.
Napoleon I, *Maxims* (1831)

5 Have I not myself known five hundred living
soldiers sabred into crows' meat for a piece
of glazed cotton which they call their flag;
which, had you sold it in any market-cross
would not have brought you above three
groschen?
Thomas Carlyle, *Sartor Resartus* (1836)

6 Hats off!
Along the street there comes
A blare of bugles, a ruffle of drums,
A flash of colour beneath the sky.
Hats off!
The flag is marching by.
Henry Holcomb Bennett, 'The Flag Goes
By'

7 I was very proud of having unfurled the
colours of my regiment before the French for
the first time and cheered loudly with the
rest when I saw them run.
George Bell, *Rough Notes by an Old Soldier*
(1867)

8 'Shoot if you must, this old grey head,
But spare your country's flag,' she said.
John Greenleaf Whittier, 'Barbara Frietchie'
(1863)

9 Swift blazing flag of the regiment,
Eagle with crest of red and gold,
These men were born to drill and die.
Stephen Crane, 'Do Not Weep', *War is
Kind* (1899)

10 A man is not a soldier until he is no longer
homesick, until he considers his regiment's
colours as he would his village steeple; until
he loves his colours, and is ready to put

hand to sword every time the honour of the
regiment is attacked.
Marshall T.R. Bugeaud, quoted in Thomas,
Les Transformations de l'Armée Française
(1887)

11 My good mother,
Be proud: I carry the flag.
Rainer Maria Rilke, 'The Lay of the Love
and Death of Cornet Christopher
Rilke', (1904)

278 COMMANDERS

1 It is the part of a good general to talk of
success, not of failure.
Sophocles, *Oedipus Coloneus*, I

2 A worthy commander is one that accounts
learning the nourishment of military virtue,
and lays that as his first foundation.
Sir Thomas Overbury, *Characters* (1614)

3 No man is fit to command another that
cannot command himself.
William Penn, *No Cross, No Crown* (1669)

4 A perfect general, like Plato's republic, is a
figment of the imagination.
King Frederick II of Prussia, *Instructions for
his Generals* (1747)

5 He who hazards nothing, gains nothing.
Napoleon I, quoted in de Liancourt,
*Political Aphorisms of the Emperor
Napoleon* (1848)

6 No army can be efficient unless it be a unit
for action, and the power must come from
above, not below.
General William T. Sherman, letter to
General Ulysses S. Grant, 1864

7 The absence of a commanding officer at the
front means that soldiers do not go into
battle but into a slaughterhouse.
V.I. Lenin, speech to 8th Communist Party
Congress, 1919

8 A competent leader can get efficient service
from poor troops; while, on the contrary, an
incapable leader can demoralise the best of
troops.
General John J. Pershing, *My Experiences in
the World War* (1931)

9 There is an intensity required in the
commander who can insist upon and obtain

that most difficult of all results, synchrony
— the arrival of this or that force exactly at
the place and moment where it is desired to
act.
Hilaire Belloc, *The Tactics and Strategy of
the Great Duke of Marlborough* (1933)

10 This is the first and true function of the
leader, never to think the battle or the cause
lost.
Field Marshal Earl Wavell, Lees-Knowles
Lecture, Cambridge, 1939

11 In war, the only sure defense is offense: and
the efficiency of offense depends on the
warlike souls of those conducting it.
General George S. Patton, *Military Review*
(1948)

12 No man rose to military greatness who
could not convince his troops that he put
them first, above all else.
General Maxwell D. Taylor, *Military
Review* (1953)

13 Men think as their leaders think.
General Charles P. Summerall, *Army
Information Digest* (1954)

14 Probably one of the greatest assets a
commander can have is the ability to radiate
confidence in the plan and operations even
(perhaps specially) when inwardly he is not
too sure about the outcome.
Field Marshal Viscount Montgomery,
Memoirs (1958)

15 The good general is the one who wins his
battles with the fewest possible casualties.
Field Marshal Viscount Montgomery,
Memoirs (1958)

16 The Commander should ensure that his
troops shall see him.
Field Marshal Earl Alexander, *Memoirs*
(1962)

17 The ability to evaluate the situation
objectively has always been a sign of true
leadership.
General Frido von Senger und Etterlin,
Neither Fear nor Hope (1964)

18 A general needs luck, and without luck his
troops will lose confidence in him.
Philip Mason, *A Shaft of Sunlight* (1978)

19 Generals tend to win their reputations at the
cost of other men's lives.
Ronald Lewin, *The Chief* (1980)

279 COMRADESHIP

See also 291. ESPRIT DE CORPS

1 My faithful friends, if we remain inseparable
in our love, we shall be invincible.
Saint Louis, address to French Crusaders,
1248

2 And much more I am soryar for my good
knyghtes losse than for the losse of my fayre
quene; for quenys I myght have inow, but
such a felyship of good knyghtes shall never
be togydirs in no company.
Thomas Malory, *The Most Piteous Tale of
the Morte Arthur Sanz Guerdon* (1470)

3 I looked alongst the line; it was enough to
assure me. The steady determined scowl of
my companions assured my heart and gave
me determination.
Anon. soldier of the Peninsular War, quoted
in Richardson, *Fighting Spirit: Psychological
Factors in War* (1978)

4 Don't think of yourself, think of your
comrades; they will think of you. Perish
yourself, but save your comrades.
General Mikhail Ivanovich Dragomirov,
Notes for Soldiers (1890)

5 East and west on fields forgotten
Bleach the bones of comrades slain,
Lovely lads are dead and rotten;
None that go return again.
A.E. Housman, *A Shropshire Lad* (1896)

6 One of most pathetic features of the war is
this continual forming of real friendships
which last a week or two, or even months,
and are suddenly shattered for ever by death
or division.
Captain Edwin Campion Vaughan, Diary,
1917

7 O beautiful men, O men I loved,
O whither are you gone, my company?
Herbert Read, 'My Company', *Naked
Warriors* (1919)

8 All things were bearable if one bore them
'with the lads.' Battles would have become
terrible beyond endurance, if pride did not
make a man endure what his comrades
endured.
Charles Carrington, *A Subaltern's War*
(1929)

9 No one who saw the 14th Army in action,
above all, no one who saw its dead in the

field of battle, the black and the white and the brown and the yellow lying together in their indistinguishable blood in the rich soil of Burma, can ever doubt there is a brotherhood of man; or fail to cry, What *is* Man, that he can give so much for war, so little for peace?
John Masters, *The Road Past Mandalay* (1961)

10 Courage is communicable from one tto another. In company, the dangers are shared, but when alone, amid darkness and death, there is no sharing.
George Coppard, *With a Machine Gun to Cambrai* (1969)

11 Numberless soldiers have died, more or less willingly, not for country or honour or religious faith or for any other abstract good, but because they realised that by fleeing their posts and rescuing themselves, they would expose their companions to greater danger.
J. Glenn Gray, *The Warriors; Reflections on Men in Battle* (1970)

12 Whatever first drove us all to war, it was friendship, trust and loyalty to one another that kept us at it long after we might have preferred to be elsewhere.
Patrick Davis, *A Child at Arms* (1971)

13 Any man in combat who lacks comrades who will die for him, or for whom he is willing to die, is not a man at all. He is truly damned.
William Manchester, *Goodbye, Darkness* (1981)

14 Those men on the line were my family, my home. They were closer to me than I can say, closer than any friends had been or ever would be.
William Manchester, *Goodbye, Darkness* (1981)

but whosoever shall smite thee on thy right cheek, turn to him the other also.
Matthew 5:39

3 The peace of the man who has forsworn the use of the bullet seems to me not quite peace, but canting impotence.
Ralph Waldo Emerson, *Essays* (1839)

4 Rendering oneself unarmed when one had been the best-armed, out of a height of feeling — that is the means to real peace, which must always rest in the peace of the mind.
Friedrich Nietzsche, *The Wanderer and his Shadow* (1880)

5 War hath no fury like a non-combatant.
C.E. Montague, *Disenchantment* (1922)

6 Some time they'll give a war and nobody will come.
Carl Sandburg, *The People, Yes* (1936)

7 It is open to a war resister to judge between the combatants and wish success to the one who has justice on his side. By so judging he is more likely to bring peace between the two than by remaining a mere spectator.
M.K. Gandhi, *Non-Violence in Peace and War* (1948)

8 One of the hardest tests of a man's moral courage is his ability to face the disapproval even of his friends for an action which strikes at all the traditions of his class but which nevertheless he feels compelled to take in order to be at ease with his own conscience.
Compton Mackenzie, *On Moral Courage* (1962)

9 The noble art of losing face
may one day save the human race
and turn into eternal merit
what weaker minds would call disgrace.
Piet Hein, 'Losing Face', *Grooks* (1966)

280 CONSCIENTIOUS OBJECTORS

1 If one has a free choice and can live undisturbed, it is sheer folly to go to war.
Pericles, on the outbreak of the Peloponnesian War, 431 BC, quoted in Thucydides, *Peloponnesian Wars*, II, 61

2 But I say unto you, That ye resist not evil;

281 CONSCRIPTION

1 The individual who refuses to defend his rights when called by his government deserves to be a slave, and must be punished as an enemy of his country and a friend to her foe.
General Andrew Jackson, proclamation to the people of Mobile, 1814

2   Our Jimmy has gone for to live in a tent
    They have grafted him into the army;
    He finally pucker'd up courage and went.
    When they grafted him into the army.
    Henry C. Work, *Grafted into the Army*
    (1862)

3   Back to the army again, sergeant,
    Back to the army again,
    Out o' the cold an' the rain.
    Rudyard Kipling, 'Back to the Army Again',
    Barrack-Room Ballads (1892)

4   England expects and needs that every
    able-bodied man should do his duty, and
    fulfil honourably the responsibilities of
    citizenship, instead of only thinking about
    the privileges.
    A.J. Dawson, *The Frontiersman's Pocket
    Book* (1908)

5   Camp life, and sleeping eight in a tent, is to
    redeem us from the pleasant vices of the
    past, and the Golden Age is to come with
    the ability of every ploughboy to handle a
    rifle.
    J.A. Farrar, *Invasion and Conscription*
    (1909)

6   There should be compulsory military service
    in order that all men may have the chance,
    which otherwise they would probably avoid,
    of developing true manhood.
    Captain Sir Basil Liddell Hart,
    'Autobiographical Note', 1914

7   Send for the boys of the Old Brigade,
    To keep Old England free.
    Send for me father and me mother and me
              brother,
    But for Gawd's sake don't send for me!
    Anon., song of the First World War,
    1914–18

8   That the English have no universal military
    service is one of the shortcomings of English
    culture.
    Heinrich von Treitschke, quoted in Müller,
    Heinrich von Treitschke (1915)

9   I didn't raise my son to be a soldier,
    I brought him up to be my pride and joy.
    Who dares to put a musket to his shoulder,
    To kill some other mother's darling boy.
    Anon., anti-conscription song, 1916

10 National Service was like a very long, very
    tedious journey on the Inner Circle.
    David Lodge, *Ginger, You're Barmy* (1962)

11 Conscription, compulsory education and
    the right to vote formed the pillars of the
    democratic state.
    H. Gollwitzer, *Europe in the Age of
    Imperialism 1880–1914* (1969)

12 For the Regular Army, National Service
    meant recurring floods of reluctant youths
    on its doorstep every fortnight, twenty-three
    times a year, all to be documented, kitted,
    drilled and transformed into something
    resembling soldiers.
    J.M. Brereton, *The British Soldier* (1986)

13 Traditionally, the people of Great Britain
    have entertained a profound revulsion
    toward the idea of maintaining large
    conscript armed forces.
    Trevor Royle, *The Best Years of Their Lives:
    The National Service Experience 1945–1963*
    (1986)

---

282 COURAGE
   See also 275. BRAVERY

---

1   They are surely to be esteemed the bravest
    spirits who, having the clearest sense of both
    the pains and pleasures of life, do not on
    that account shrink from danger,
    Thucydides, *Peloponnesian Wars*, trs.
    Jowett

2   The less they spared themselves in battle,
    the safer they would be.
    Salust, *Bellum Iugurthinum*

3   Thought shall be harder, heart the keener,
    courage the greater as our might lessens.
    Byrhtnoth, exhortation to his troops at the
    Battle of Maldon, 927, quoted in *The
    Battle of Maldon* (c.1000)

4   If you bear yourselves valiantly you will
    obtain victory, honour and riches. If not,
    you will be ruthlessly butchered, or else led
    ignominiously captive into the hands of
    ruthless enemies.
    William of Poitiers, order to his men before
    the Battle of Hastings, 1066, quoted in
    *Gesta Willielmi Ducis Normannorum et
    Regis Anglorum* (c. 1071), trs. Douglas and
    Greenaway

5 Come partner, there is nothing like having good Courage.
Matthew Bishop, *Life and Adventures* (1744)

6 To appear the more warlike, you should ride with your sword drawn; but take care you do not cut your horse's ear off.
Francis Grose, *Advice to the Officers of the British Army* (1782)

7 Courage is like love: it must have hope for nourishment.
Napoleon I, *Maxims* (1804–15)

8 As to moral courage, I have rarely met with two o'clock in the morning courage: I mean unprepared courage.
Napoleon I, quoted in Las Cases, *Mémorial de Sainte Hélène* (1815)

9 Any coward can fight a battle when he's sure of winning; but give me the man who has the pluck to fight when he's sure of losing.
George Eliot, *Janet's Repentance* (1857)

10 Courage is resistance to fear, mastery of fear — not absence of fear.
Mark Twain, *Pudd'nhead Wilson* (1894)

11 The paradox of courage is that a man must be a little careless of his life even in order to keep it.
G.K. Chesterton, *All Things Considered* (1908)

12 Courage is the thing. All goes if courage goes.
Sir James Barrie, Rectorial Address at St Andrews, 1922

13 Untutored courage is useless in the face of educated bullets.
General George S. Patton, *Cavalry Journal* (1922)

14 The war has given birth to that weird courage which inspires a man to great bravery even when he knows no reward will or can possibly accrue: the daring is reasoned, like the clarity of a man meeting death open-eyed and never wavering even an eye-lid.
Hugh Quigley, *Passchendaele and the Somme* (1928)

15 Courage is will-power, whereof no man has an unlimited stock; and when in war it is used up, he is finished.
Lord Moran, *The Anatomy of Courage* (1945)

16 A man of character in peace is a man of courage in war.
Lord Moran, *The Anatomy of Courage* (1945)

283 COWARDICE

1 How cowardly to escape oneself and leave a friend in the lurch!
Euripides, *Iphigenia in Tauris*

2 When soldiers run away in war they never blame themselves: they blame their general or their fellow-soldiers.
Demosthenes, *Third Olynthiac*

3 The soldiers saw that those who ran away were either captured or slain, while the bravest were the safest.
Sallust, *Bellum Iurgurthinum*

4 Cowards die many times before their
        deaths;
The valiant never taste of death but once.
William Shakespeare, *Julius Caesar*, II,ii

5 Few cowards always know the full extent of their fear.
Duc de La Rochefoucauld, *Maxims* (1678), trs. Tancock

6 Many would be cowards if they had courage enough.
Thomas Fuller, *Gnomologia* (1732)

7 Tears are no proof of cowardice, Trim; I drop them oft-times myself.
Laurence Sterne, *The Life and Opinions of Tristram Shandy, Gentleman* (1760–67)

8 For he who fights and runs away
May live to fight another day;
But he who is in battle slain
Can never rise and fight again.
Oliver Goldsmith, 'The Art of Poetry in a New Place'

9 Coward's funerals when they come,
Are not wept so well at home.
A.E. Housman, *A Shropshire Lad* (1896)

10 Coward: one who in a perilous emergency thinks with his legs.
Ambrose Bierce, *The Devil's Dictionary* (1906)

11 It seems that when a man turns his back to

the front he is on the road to panic and when his mind and body have reached breaking-point the road is not so long.
Major Desmond Allhausen, Diary, 1917, quoted in Warner, *Passchendaele* (1987)

12 Cowardice, as distinguished from panic, is almost always simply a lack of ability to suspend the functioning of the imagination.
Ernest Hemingway, introduction to *Men at War* (1942)

13 Cowards are those who let their timidity get the better of their manhood.
General George S. Patton, letter to his son, 1944

14 Most of the time the word *realism* is a polite translation of the word *cowardice*.
Jean Dutourd, *Taxis of the Marne* (1957)

15 Let us be wary of ready-made ideas about cowardice and courage: the same burden weighs infinitely more heavily on some shoulders than on others.
François Mauriac, *Second Thoughts* (1961), trs. Foulke

DECORATIONS
See 305. MEDALS

284 DEMOBILIZATION

1 Scatter thou the people that delight in war.
Psalms, 68:30

2 Deliver to the army this news of peace; let them have pay and part.
William Shakespeare, *Henry IV* Part 2, IV, ii

3 No sooner had I got my discharge in my pocket than I felt I was a new man; I was once more free; I actually thought I stood a few inches higher, as I stretched myself like one who has just laid down a heavy load.
Alexander Alexander, *Life* (1830)

4 For the first time since I had been a shepherd lad on Blandford Downs, I saw myself in plain clothes, and with liberty to go and come where I like.
Rifleman John Harris, *Recollections* (1848)

5 Had I my life in front of me instead of behind, I would start again, just as I did when I was a lad of eighteen and desire nothing better than to live those happy twenty-five years over again in the ranks of the Old 14th as a King's Hussar.
Edwin Mole, *A King's Hussar* (1893)

6 'Twas quite nice to be a soldier for a year,
But it's nicer to feel free once more.
There's quite enough depravity, pain and
  fear
In life's remorseless chore.
Alfred Lichtenstein, 'Abschied June 1914'

7 It was not easy to begin again, to take thought for the morrow when you had not expected to be alive for it.
Charles Carrington, *A Subaltern's War* (1929)

8 We were all accustomed to the war-time view, that the sole qualification for peace-time employment would be a good record of service in the field, that we expected our scars and our commanding officers' testimonials to get us whatever we wanted.
Robert Graves, *Goodbye to All That* (1929)

9 How could I begin my life all over again when I had no conviction about anything except that the War was a dirty trick which had been played on me and my generation?
Siegfried Sassoon, *Sherston's Progress* (1936)

10 America fought the war like a football game, after which the winner leaves the field and celebrates.
General Albert C. Wedemeyer, 1945, on the demobilisation of the US Army troops at the end of the Second World War

11 Looking back at those firm ranks as they marched into billets, I found that a body of men had become so much a part of me that its disintegration would tear away something I cared for more dearly than I could have believed. I was it and it was I.
Guy Chapman, *Vain Glory* (1968)

12 The youth had become a man but with only the capabilities of a youth to meet adult realities in civvy street. Although an expert machine gunner, I was a numbskull so far as any trade or craft was concerned.
George Coppard, *With a Machine Gun to Cambrai* (1969)

285 DESERTERS

1 Not only no traitor, but no deserter even,
has found a place in my camp.
Cicero, *In Verrem*, II, 1

2 An adversary is more hurt by desertion than
by slaughter.
Vegetius, *De Re Militari* (378)

3 That, sire, which serves and seeks for gain,
And follows but for form,
Will pack when it begins to rain,
And leave thee in the storm.
William Shakespeare, *King Lear*, II, iv

4 Deserted at his utmost Need,
By those his former Bounty fed:
On the bare Earth expos'd he lyes,
With not a Friend to close his Eyes.
John Dryden, 'Alexander's Feast' (1697)

5 The first time I deserted I thought myself
free;
Informed on by my comrades a deserter to
be,
I was soon followed after and brought back
with speed,
I was handcuffed and shackled,
heavy-ironed indeed.
Anon., song of *c.*1819, 'The New Deserter'

6 A soldier who offers to quit his rank, or
offers to flag, is to be instantly put to death
by the officer who commands that platoon,
or by the officer or sergeant in the rear of
that platoon; a soldier does not deserve to
live who won't fight for his king and
country.
General James Wolfe, Order to the 20th
Foot, Canterbury, 1755

7 He had deserted in a blink of fine weather
between the rains that splashed the glutted
rat-runs of the front. He had done it quickly
and easily . . . he had just turned and
walked back.
Lewis Grassic Gibbon, *Sunset Song* (1932)

8 The vast majority of deserters and those
going AWOL during the Vietnam era, as in
previous wars absented themselves not for
political reasons but because of personal or
financial problems or inability to adjust to
military life.
Guenter Lewy, *America in Vietnam*
(1978)

286 DICTATORS

1 Death is a softer thing by far than tyranny.
Aeschylus, *Agamemnon*

2 The strongest shield and safeguard for all
men, especially for the masses against
tyrants, is mistrust.
Demosthenes, *Philippica*, II, 24

3 *Minimi jura, quotiens gliscat potestas, nec
utendum imperio, ubi legibus agi possit.*
Rights are invariably abridged as despotism
increases; nor ought we to fall back on
imperial authority when we can have
recourse to the laws.
Tacitus, *Annals*, III, 69

4 Kings will be tyrants from policy, when
subjects are rebels from principle.
Edmund Burke, *Reflections on the
Revolution in France* (1790)

5 Among those who dislike oppression are
many who like to oppress.
Napoleon I, *Political Aphorisms* (1848)

6 In so far as we educate the people to fight
against the delirium of democracy and bring
it again to the recognition of the necessity of
authority and leadership, we tear it away
from the nonsense of parliamentarianism.
Adolf Hitler, speech at Nuremburg, 1928

7 The tyrant claims freedom to kill freedom,
And yet to keep it for himself.
Rabindranath Tagore, *Fireflies* (1928)

8 He knew human folly like the back of his
hand,
And was greatly interested in armies and
fleets;
When he laughed, respectable senators burst
with laughter,
And when he cried the little children died in
the streets.
W.H. Auden, 'Epitaph on a Tryant'

9 Dictators ride to and fro upon tigers which
they dare not dismount. And the tigers are
getting hungry.
Winston S. Churchill, *While England Slept*
(1936)

10 And the little screaming fact that sounds
through all history: repression works only to
strengthen and knit the oppressed.
John Steinbeck, *The Grapes of Wrath* (1939)

11 So long as men worship the Caesars and
Napoleons, Caesars and Napoleons will
duly rise and make them miserable.
Aldous Huxley, *Ends and Means* (1937)

12 A police state finds it cannot command the
grain to grow.
John F. Kennedy, State of the Union
Message, 1963

13 The benevolent despot who sees himself as a
shepherd of the people still demands from
others the obedience of sheep.
Eric Hoffer, *The Ordeal of Change* (1964)

287 DISCIPLINE

1 It is a bad soldier who grumbles when
following his commander.
Seneca, *Epistulae ad Lucilium Epis*, CVII

2 You do well to comprehend that good order
and military discipline are the chief
essentials in an army.
Duke of Marlborough, letter to Colonel
William Cadogan, 1703

3 The finest requisite for the real business of
fighting is only to be attained by a constant
course of discipline and service.
General George Washington, speech to
Congress, 1780

4 The chief virtues of a soldier are constancy
and discipline. Valour comes only in second
place.
Napoleon I, proclamation to his army, 1799

5 The qualities which, commonly, make an
army formidable, are long habits of
regularity, great exactness of discipline and
great confidence in the commander.
Samuel Johnson, quoted in *Works*, vol. VI
(1825)

6 Too many courts-martial in any command
are evidence of poor discipline and
inefficient officers.
General William T. Sherman, letter to
General Ulysses S. Grant, 1864

7 The training of the rank and file for war
consists in teaching three things — drill,
musketry (or gunnery) and discipline. But of
these essentials, discipline is by far the most
essential, for discipline, even more than drill
or musketry, leads to victory, and discipline
cannot be imparted in small doses.
Field Marshal Earl Roberts, *Fallacies and
Facts* (1911)

8 The sterner the discipline, the better the
soldier, the better the army.
Stephen Graham, *A Private in the Guards*
(1919)

9 No sane man is unafraid in battle, but
discipline produces in him a form of
vicarious courage.
General George S. Patton, *War As I Knew It*
(1947)

10 Discipline is based on pride in the profession
of arms, a meticulous attention to details,
and in mutual respect and confidence.
Discipline must be a habit so ingrained that
it is stronger than the excitement of battle or
the fear of death.
General George S. Patton, *War As I Knew It*
(1947)

11 Discipline is founded on the good will of all
ranks to do their best at all times.
Lt-Gen Sir Francis Tuker, *While Memory
Serves* (1950)

12 I am confident that an Army of stray
individuals held together by a sound
discipline based on respect for personal
initiative and rights and dignity of the
individual will never fail this nation in time
of need.
General J. Lawton Collins, *Military Review*
(1951)

13 We very soon learned that strict discipline in
battle and bivouac was vital, not only for
success, but for survival.
Field Marshal Viscount Slim, *Defeat into
Victory* (1956)

14 One's natural instinct when shooting starts
is to lie in a ditch and stay there until it is all
over; and it is only through discipline and
training that one can make oneself get out
and go forward.
Brigadier Sir John Smyth, *Sandhurst* (1961)

15 Discipline in its finest sense is the cheerful
obedience of orders.
Byron Farwell, *The Gurkhas* (1984)

288 DRILL

1 Troops who march in an irregular and disorderly manner are always in great danger of being defeated.
Vegetius, *De Re Militari* (378)

2 Drill is necessary to make the soldier steady and skilful, although it does not warrant exclusive attention.
Marshal Maurice de Saxe, *Mes Rêveries* (1757)

3 Those who have never seen two well-trained armies drawn up for battle can have no idea of the beauty and brilliance of the display.
Voltaire, *Candide* (1759), trs. Butt

4 What makes a regiment of soldiers a more noble object of view than the same mass of mob? Their arms, their dress, their banners and their art and artificial symmetry of their position and movements.
Lord Byron, letter to John Murray, 1821

5 I would rather be shelled for an hour than drilled for an hour.
Raymond Asquith, letter to his wife, 1916

6 There is a religious or poetic element in perfecting even one's dressing by the right.
Edmund Blunden, *Undertones of War* (1928)

7 At the hoary bellow of a single man
The ragged line of disparate identity
Springs to spontaneous activity:
Rhythmic rocking, hobnails knocking,
Bodies to the sacrifice take their advice.
Timothy Corsellis, 'Drill', 1940, quoted in Blythe, *Components of the Scene* (1966)

8 It has been found from experience that units which are good at drill and which are well turned out rarely fail in battle.
Brigadier Sir John Smyth, *Before the Dawn* (1957)

9 No one who has participated in it or seen it well done should doubt the inspiration of ceremonial drill.
Field Marshal Earl Wavell, *Soldiers and Soldiering* (1953)

10 We found it a great mistake to belittle the importance of smartness in turn-out, alertness of carriage, cleanliness of person, saluting or precision of movement, and to dismiss them as naive, unintelligent, parade-ground stuff.
Field Marshal Viscount Slim, *Defeat into Victory* (1956)

11 It is astonishing how much of the business of warfare can be carried on by men who act as automatons, behaving almost as mechanically as the machines they operate.
J. Glenn Gray, *The Warriors: Reflections on Men in Battle* (1970)

12 Drill as the means to an end is indispensable to every army. It cannot be replaced by individual training nor by sporting instinct.
Herbert Sulzbach, *With the German Guns* (1981)

289 DRINKING

1 A capitayn sholde lyve in sobrenesse.
Geoffrey Chaucer, *The Canterbury Tales*, 'The Pardoner's Tale' (*c*.1387)

2 And let me the canakin clink:
A soldier's a man;
A life's but a span;
Why then let a soldier drink.
William Shakespeare, *Othello*, II, iii

3 Who can look for modestie and sobrietie in the souldiers, where the captaine is given to wine or women, and spendeth his time in riot and excesse?
Mathew Sutcliffe, *The Practice, Proceedings and Lawes of Armes* (1593)

4 You stink of brandy and tobacco, most soldier-like.
William Congreve, *The Old Bachelor*, III (1693)

5 Drinking is the soldier's pleasure.
John Dryden, 'Alexander's Feast' (1697)

6 We bravely fought and conquered
At Ross and Wexford town;
And if we failed to keep them,
'Twas drink that brought us down.
Robert D. Joyce, attr., 'The Boys of Wexford'

7 The English soldiers are fellows who have

all enlisted for drink — that is the plain fact
— they have all enlisted for drink.
Duke of Wellington, 1831, quoted in
Stanhope, *Notes of Conversations with the
Duke of Wellington* (1888)

8   The State offered every inducement in the
way of monotonous diet, monotonous
occupation, climatic discomfort, bad
housing and abundant alcohol that could
lure men to drink; and then deplored the
drunkenness of the Army.
Sir John Fortescue, *A Short Account of
Canteens in the British Army* (1928)

9   Sometimes the Russians supplied vodka to
their storm battalions, and the night before
the attack we could hear them roaring like
devils.
Maj-Gen F.W. von Mallenthin, *Panzer
Battles* (1955)

10 It's the easiest way to make heroes. Vodka
purges the brain and expands the strength.
Guy Sajer, *Forgotten Soldier* (1971)

11 To the civilian drunkenness was merely a
vice; to the military it was a crime,
punishable with the lash.
J.M. Brereton, *The British Soldier* (1986)

---

290 ENEMY

1   It is right to be taught, even by an enemy.
Ovid, *Metamorphoses*, IV

2   A dead enemy always smells good.
Alus Vitellius, on the battlefield of
Beariacum, AD 69

3   The last enemy that shall be destroyed is
death.
1 Corinthians 25:26

4   'Tis best to weigh
The enemy more mighty than he seems.
William Shakespeare, *Henry V*, II, iv

5   If we are bound to forgive an enemy, we are
not bound to trust him.
Thomas Fuller, *Gnomologia* (1732)

6   Respect was mingled with surprise,
And the stern joy which warriors feel
In foemen worthy of their steel.
Sir Walter Scott, *The Lady of the
Lake*, X (1810)

7   He who lives by fighting with an enemy has
an interest in the preservation of the enemy's
life.
Friedrich Nietzsche, *Human, All Too
Human* (1878)

8   So 'ere's to you, Fuzzy-Wuzzy, at your
        'ome in the Soudan;
You're a pore benighted 'eathen, but a
        first-class fightin' man.
Rudyard Kipling, 'Fuzzy-Wuzzy',
Barrack-Room Ballads (1892)

9   I am the enemy you killed, my friend.
Wilfred Owen, 'Strange Meeting' (1918)

10 I hate them but do not despise them; I am
proud of my enemy, essential for a fighting
man.
Ernst Junger, *The Storm of Steel* (1929)

11 That duty which compels you to kill a man
does not compel you to hate him.
Charles Carrington, *A Subaltern's War*
(1929)

12 When you encounter the enemy after
landing, think of yourself as the avenger
come at last face to face with your father's
murderer.
Colonel Masonobu Tsuji, order to his
troops before the Japanese invasion of
Malaya, 1941

13 The basic aim of a nation at war in
establishing an image of the enemy is to
distinguish as sharply as possible the act of
killing from the act of murder by making the
former into one deserving all honour and
praise.
J. Glenn Gray, *The Warriors: Reflections on
Men in Battle* (1970)

---

291 ESPRIT DE CORPS
See also 279. COMRADESHIP

---

1   The company of just and righteous men is
better than wealth and a rich estate.
Euripides, *Aegeus*, 7

2   Such a fellowship of good knights shall
never be together in no company.
Thomas Malory, *Morte d'Arthur*, XX, 9
(1485)

3 To insure victory the troops must have confidence in themselves as well as in their commanders.
Niccolo Machiavelli, *Discorsi*, XXXIII (1531)

4 We would not die in that man's company That fears his fellowship to die with us.
William Shakespeare, *Henry V*, IV, iii

5 Ours was an esprit de corps — a buoyancy of feeling animating all which nothing could quall. We were alike ready for the field or for frolic, and when not engaged in the one, went headlong into the other.
John Kincaid, of the British Army in the Peninsular War, 1811, quoted in *Random Shots from a Rifleman* (1835)

6 It is not set speeches at the moment of battle that renders soldiers brave. The veteran scarcely listens to them, and the recruit forgets them at the first discharge. If discourses and harangues are useful, it is during the campaign; to do away with unfavourable impressions, to correct false reports, to keep alive a proper spirit in camp, and to furnish materials and amusement for the bivouac.
Napoleon I, quoted in Wavell, *The Good Soldier* (1948)

7 Discipline cannot be secured or created in a day. It is an institution, a tradition.
Charles Ardant du Picq, *Battle Studies* (1870)

8 Unhappy is that man who is not sensitive to crowd-emotion, for he bears the weight of war on his shoulders alone. To such a man war is indeed a nightmare.
Charles Carrington, *A Subaltern's War* (1929)

9 Living in an atmosphere of soldierly duty and esprit de corps permeates the soul, where drill merely attunes the muscles.
Captain Sir Basil Liddell Hart, *Thoughts on War* (1944)

10 (Of the British Army) Men of our race grow in adversity; grave news binds them into one happy company.
Lord Moran, *The Anatomy of Courage* (1945)

11 By the spirit of the group and one's integration with it one was largely relieved by the burden of individual life.
Richard Rumbold, *My Father's Son* (1949)

12 There is no substitution for the personal sharing of dangers.
Maxwell Taylor, *The Uncertain Trumpet* (1959)

292 FEAR

1 The fear of war is worse than fear itself.
Seneca, *Hercules Furens*

2 The thing in the world I am most afraid of is fear.
Michel de Montaigne, *Essays*, I (1580)

3 The first duty of a man is still that of subduing *fear.*
Thomas Carlyle, *On Heroes, Hero-Worship and the Heroic in History* (1841)

4 It is not the fear of death — that is past, thank God — but I fear defeat, and its consequences.
General Charles George Gordon, Khartoum Journal, 1883

5 The best antidote to fear is food.
Captain Edwin Campion Vaughan, Diary, 1917

6 The only thing I'm not certain about is whether I may get the wind up and show it. I'm afraid of being afraid.
Captain J.E.H. Neville, letter to his father, 1917, quoted in *The War Letters of a Light Infantryman* (1930)

7 Fear is the worst of the horrors of war. Fear is that which degrades, which breeds cruelty, envy and malice; and fear is the enemy in war.
Charles Carrington, *A Subaltern's War* (1929)

8 The real enemy was Terror, and all this heel-clicking, saluting, bright brass and polish were our charms and incantations for keeping him at bay.
Alan Hanbury-Sparrow, *The Land-Locked Lake* (1932)

9 I am afraid, to a greater or lesser degree, in every battle. I cannot say that I have overcome my fear, but rather that my fear has not yet overcome me.
Jack Belden, *Still Time to Die* (1944)

10 But I was a coward; and the thing I feared

more than anything in the world was to
break up in battle and give way to that
cowardice.
John Watney, *The Enemy Within* (1946)

11 The worst part of battle was wondering how
you were going to behave in front of other
people.
Raleigh Trevelyan, *The Fortress* (1956)

12 The state of mind that action induces
primarily and superficially is fear, with
peaks of almost hysterical tension. Fear
becomes commonplace — like death, an
accepted everyday, ever-present condition.
Neil McCallum, *Journey With A Pistol*
(1959)

13 It is only the complete absence of an enemy
that makes a soldier feel heroic.
Peter Cochrane, *Charlie Company* (1977)

14 Fear is the common bond between fighting
men.
Richard Holmes, *Firing Line* (1985)

FLAG
See 277. COLOURS

FOOD
See 318. RATIONS

---

293 FREEDOM

1 A! fredome is a noble thing!
Fredome mayse man to haiff liking.
Fredome al solace to man giffis;
He levys at eas that frely levys.
John Barbour, *The Bruce* (c.1375)

2 Those who expect to reap the blessings of
freedom must, like men, undergo the fatigue
of supporting it.
Thomas Paine, *The Crisis*, IV (1777)

3 For Freedom's battle once begun,
Bequeathed by bleeding Sire to Son,
Though baffled oft is ever won.
Lord Byron, *The Giaour* (1813)

4 If a nation expects to be ignorant and free,

in a state of civilisation, it expects what
never was and never will be.
Thomas Jefferson, letter to Colonel Charles
Yancey, 1816

5 Yet, Freedom! yet thy banner, torn, but
flying,
Streams like the thunder-storm *against* the
wind.
Lord Byron, *Childe Harold's Pilgrimage*, IV
(1818)

6 I gave my life for freedom — This I know:
For those who bade me fight had told me so.
William Norman Ewer, 'Five Souls' (1917)

7 What stands if Freedom fall?
Rudyard Kipling, 'For all we have and are',
Verse (1940)

8 If a nation values anything more than
freedom, it will lose its freedom; and the
irony of it is that if it is comfort or money it
values more, it will lose that too.
Somerset Maugham, *Strictly Personal*
(1941)

9 National strength lies only in the hearts and
spirits of men.
S.L.A. Marshall, *Men Against Fire* (1947)

10 Let every nation know, whether it wishes us
well or ill, that we shall pay any price, bear
any burden, meet any hardship, support any
friend, oppose any foe, to assure the survival
and success of liberty.
John F. Kennedy, Inaugural Address as
President, 1961

---

294 FRONT LINE

1 A battlefield is at once the playroom of all
the gods and the dancehall of all the furies.
Jean Paul Richter, *Titan* (1803)

2 Close combat, man to man, is plainly to be
regarded as the real basis of combat.
Karl von Clausewitz, *On War* (1832)

3 The front was a moloch that consumed
bodies, but souls were often tempered in its
fire.
Captain Sir Basil Liddell Hart, *Lawrence*
(1934)

4 The man in the firing-line is like a man in

the depth of a forest: he cannot see the forest for trees.
Douglas Hallam, 'Quinn's and Courtney's'. *Blackwood's Magazine* 1939

5 It is always a minority that occupies the front line.
Maj-Gen Orde Wingate, Order of the Day, 1942

6 Combat soldiers are an exclusive set, and if they want to be that way, it is their privilege.
William H. Mauldin, *Up Front* (1945)

7 Soldiers who are in danger feel natural and human resentment to those that aren't.
William H. Mauldin, *Up Front* (1945)

8 Many soldiers — tired by the rigidities of normal life — look back at violent moments of their war experiences despite the hunger and terror, as the monumental culminating experiences of their lives.
J.A.M. Meerloo, *Mental Seduction and Menticide* (1957)

9 I have always regarded the forward area of the battlefield as the most exclusive club in the world, inhabited by the cream of the nation's manhood — the men who actually do the fighting.
Lt-Gen Sir Brian Horrocks, *A Full Life* (1960)

10 In all great battles involving great masses of men, the individual soldier knows nothing of what is happening a few yards beyond him.
Leonard Cottrell, *Enemy of Rome* (1960)

11 (Of the Western Front, 1917) Life was very simple at the front, you just served.
P.J. Campbell, *In the Cannon's Mouth* (1977)

12 Few young soldiers in any army, facing battle for the first time, are prepared for the noise, muddle, blood, horror and their own fear.
Byron Farwell, *The Gurkhas* (1984)

GAMES
See 323. SPORT

295 HONOUR

1 When dangers are great, there the greatest honours are to be won by men and the states.
Thucydides, *Peloponnesian Wars*, II, 42

2 Honour is a mere scutcheon.
William Shakespeare, *Henry IV* Part 1, V, i

3 Mine honour is my life; both grow in one; Take honour from me and my life is done.
William Shakespeare, *Richard II*, I, i

4 He that is valiant and dares fight, though drubbed, can lose no honour by't.
Samuel Butler, *Hudibras* (1663)

5 In time of war most men will face just enough danger to keep their honour intact, but few are prepared to go on doing so long enough to ensure the success of the enterprise for which the danger is being faced.
Duc de La Rochefoucauld, *Maxims* (1678), trs. Tancock

6 Glory is the sodger's prize; The sodger's wealth is honour.
Robert Burns, 'Where the Wild War's Blast was blown'

7 Remember those who have done kindness to our race, and pay their services with thy blood, should the hour require it.
Sir Walter Scott, *A Legend of Montrose* (1819)

8 The soldier's trade, verily and essentially, is not slaying, but being slain. This, without knowing its own meaning, the world honours it for.
John Ruskin, *Unto This Last* (1895)

9 To set the cause beyond renown, To love the game beyond the prize, To honour while you strike him down, The foe that comes with fearless eyes.
Sir Henry Newbolt, 'Clifton Chapel' (1897)

10 Honour is manly decency. The shame of being found wanting in it means everything to us.
Alfred de Vigny, *The Military Necessity* (1953)

11 (Of the Spanish Foreign Legion) Battle was to be the purpose of his life; death in action his greatest honour; cowardice the ultimate disgrace.
Peter Kemp, *Mine Were of Trouble* (1957)

12 The soldier seals his devotion to his craft with his life.
Philip Mason, *A Matter of Honour* (1974)

296 LEADERSHIP

1 Pay well, command well, hang well.
Sir Ralph Hopton, *Maxims for the Management of an Army* (1643)

2 French officers will always lead, if the soldiers will follow; and English soldiers will always follow, if their officers will lead.
Samuel Johnson, *On the Bravery of the English Common Soldier* (1760)

3 The more a leader is in the habit of demanding from his men, the surer he will be that his demands are answered.
Karl von Clausewitz, *On War* (1832)

4 At the head of an army nothing is more becoming than simplicity.
Napoleon I, *Political Aphorisms* (1848)

5 No matter what may be the ability of the officer, if he loses the confidence of his troops, disaster must sooner or later ensure.
General Robert E. Lee, letter to Jefferson Davis, 1863

6 The true way to be popular with troops is not to be free and familiar with them, but to make them believe you know more than they do.
General William T. Sherman, letter to the Rt. Rev. Henry C. Lay, 1864

7 In action one may say nothing. The leader is the man who does not talk.
General Charles de Gaulle, note as a prisoner of war, 1916

8 No amount of study or learning will make a man a leader unless he has the natural qualities of one.
Field Marshal Earl Wavell, in *The Times*, 1941

9 When I make plans, always at the back of my mind is the thought that I am playing with human lives. Good chaps get killed or wounded and that is a terrible thing.
Field Marshal Earl Alexander, address to war correspondents, Italy, 1944

10 The genius of a good leader is to leave behind him a situation which common sense, without the grace of genius, can deal with successfully.
Walter Lippmann, *Roosevelt Has Gone* (1945)

11 The good company has no place for the officer who would rather be right than loved, for the time will quickly come when he walks alone, and in battle no man may succeed in solitude.
S.L.A. Marshall, *Men Against Fire* (1947)

12 All very successful commanders are prima donnas, and must be so treated.
General George S. Patton, *War As I Knew It* (1947)

13 A leader is a man who has the ability to get other people to do what they don't want to do, and like it.
Harry S. Truman, *Memoirs* (1955)

14 Men are neither lions nor sheep. It is the man who leads them who turns them into either lions or sheep.
Jean Dutourd, *Taxis of the Marne* (1957)

15 The first thing a young officer must do when he joins the Army is to fight a battle, and that battle is for the hearts of his men. If he wins that battle and subsequent similar ones, his men will follow him anywhere; if he loses it, he will never do any real good.
Field Marshal Viscount Montgomery, *Memoirs* (1958)

16 Command doth make actors of us all.
John Masters, *The Road Past Mandalay* (1961)

297 LEAVE

1 I would give all my fame for a pot of ale and safety.
William Shakespeare, *Henry V*, III, ii

2 Always ask for leave at all times and in all places, and in time you will acquire a right to it.
Francis Grose, *Advice to the Officers of the British Army* (1782)

3 We're goin' 'ome, we're goin' 'ome,
Our ship is *at* the shore,
An' you must pack your 'aversack,
For we won't come back no more.
Rudyard Kipling, 'Troopin', *Barrack-Room Ballads* (1892)

4 I'm goin' 'ome to Blighty — ain't I glad to
'ave the chance!
I'm loaded up wiv fightin', and I've 'ad my
fill o' France;
I'm feelin' so excited-like, I want to sing and
dance,

For I'm goin' 'ome to Blighty in the
      mawnin'.
Robert Service, 'Going Home', *Rhymes of a
Red-Cross Man* (1916)

5   Respite for a week or a day from the fear of
death gave absolute enjoyment for a week or
a day.
Charles Carrington, *A Subaltern's War*
(1929)

6   What is leave? — A pause that only makes
everything after it so much worse.
Erich Maria Remarque, *All Quiet on the
Western Front* (1929)

7   A certain amount of leave, although not in
any contract of service, is a recognised part
of a soldier's life.
Winston S. Churchill, memorandum to the
Minister of Pensions, 1943

8   Shall I get drunk or cut myself a piece of
      cake,
a pale Syrian with a few words of English
or the Turk who says she is a princess; she
      dances
by apparent levitation?
Keith Douglas, 'Cairo Jag', *Selected Poems*
(1943)

9   My worry was not about the war itself, but
about the possibility of its starting before I
had had my first furlough.
John Masters, *Bugles and a Tiger* (1956)

10 Delousing yourself was one of the
pleasantest prospects of going out into rest.
Charles Carrington, *Soldier from the Wars
Returning* (1965)

11 Men would dash from the front as fast as
they could, knowing that a chance shell or
bullet, a raid or a battle planned might steal
the prized leave.
Denis Winter, *Death's Men* (1978)

12 Far from giving the soldier a well-earned
respite from the pressure of war, home
leave — or even rest and recreation (R&R)
overseas — may only emphasise the physical
discomforts of the front and remind him of
what he stands to lose for ever.
Richard Holmes, *Firing Line* (1985)

298 LETTERS FROM HOME

1   Don't send a word of sorrow,

Send him a page of joy,
And don't let your teardrops
Fall upon the kisses
When you write to your soldier boy.
Anon., popular song on the Home Front,
1916, quoted in Macdonald, *Voices and
Images of the Great War* (1988)

2   Bad news from home might affect a soldier
in one of two ways. It might either drive him
to suicide (or recklessness amounting to
suicide), or else seem trivial by contrast with
present experiences and be laughed off.
Robert Graves, *Goodbye to All That* (1929)

3   Strong men often cry on post-day.
Michael Morris, *Terrorism* (1971)

4   Concern about bad news from home can
eclipse physical dangers.
Richard Holmes, *Firing Line* (1985)

299 LOOTING

1   When you plunder a countryside, let the
spoil be divided amongst your men; when
you capture new territory, cut it up into
allotments for the benefit of the soldiers.
Sun Tzu, *The Art of War*, VII, 20 (*c*.490 BC)

2   The mortal who sacks fallen cities is a fool;
Who gives the temples and the tombs, the
      hallowed places
Of the dead to desolation. His own turn will
      come.
Euripides, *The Trojan Women*, 95–7, trs.
Lattimore

3   It is a law established for all time among all
men that when a city is taken in war, the
persons and the property of the inhabitants
belong to the captors.
Xenophon, *Cyropaedia*, VII, 5, 73

4   When anyone brings home something he
has taken from the enemy with his own
hand, it affords him more pleasure and
gratification than if he were to receive many
times its value at the bidding of another.
Livy, *Histories*, V, 20

5   (Of the sack of Nineveh, 612 BC)
Take ye the spoil of silver, take the spoil of
gold: for there is none end of the store, and
glory out of all the pleasant furniture.
Nahum 2;3

6   The man who does not know how to set
places on fire, to rob churches and to usurp

their rights and to imprison the priests, is not fit to carry on war.
Honoré Bonet, *Tree of Battles* (*c.*1380), trs. Coopland

7 So many goodly cities ransacked and razed; so many nations destroyed and made desolate; so infinite millions of harmless people of all sexes, states and ages massacred, ravaged and put to the sword; and the richest, the fairest and best part of the world overturned and defaced for the traffic of pearls and pepper. O base conquest!
Michel de Montaigne, *Essays* III (1588)

8 Every day our soldiers by stealth do visit papists' houses, and constrain from them both meat and money. They give them whole great loaves and cheeses, which they triumphantly carry away upon the points of their swords.
Sergeant Nehemiah Wharton, letter from the Earl of Essex's army, 1642

9 (Of the Battle of Naseby, 1645) The whole booty of the Field fell to the soldier, which was very rich and considerable, there being amongst it, beside the riches of the court and officers, the rich plunder of Leicester.
Joshua Sprigge, *Anglia Rediviva* (1647)

10 Those possessions short-liv'd are, Into which are come by war.
Robert Herrick, *Hesperides* (1648)

11 A rapacious and licentious soldiery.
Edmund Burke, speech in Parliament on Fox's East India Bill, 1783

12 Why, they call a man a robber if 'e stuffs 'is
      marchin' clobber
With the —
Loo! loo! lulu! lulu! Loo! loo! Loot! loot!
      loot
Ow the loot!
Bloomin' loot!
Rudyard Kipling, 'Loot', *Barrack-Room Ballads of the Boer War* (1902)

---

300 LOVE

---

1 As it is base for a soldier to love; so am I in love with a base wench.
William Shakespeare, *Love's Labour Lost*, I, ii

2 He that hath wife and children hath given hostages to fortune; for they are impediments to great enterprises, either of virtue or mischief.
Francis Bacon, *Essays*, VIII (1625)

3 I know not how, but martial men are given to love. I think it is, but as they are given to wine; for perils commonly ask to be paid in pleasures.
Francis Bacon, *Essays*, X (1625)

4 Old soldiers, sweetheart, are surest, and old lovers are soundest.
John Webster, *Westward Hoe*, II, ii (1638)

5 Tell me not (Sweet) that I am unkinde, That from the Nunnerie Of thy chaste breast, and quiet minde, To Warre and Armes I flie.
Richard Lovelace, 'To Lucasta, Going to the Warres' (1649)

6 The same heat that stirs them up to Love, spurs them on to war.
George Farquhar, *The Recruiting Officer* (1706)

7 Marriage is good for nothing in the military profession.
Napoleon I, *Political Aphorisms* (1848)

8 You love us when we're heroes, home on
      leave.
Or wounded in a mentionable place.
Siegfried Sasson, 'Glory of Women', *Counter-Attack* (1918)

9 Subalterns may not marry, captains may marry, majors should marry, colonels must marry.
Anon., traditional British Army saying, *c.*1920

10 A soldier has no business to be married.
General Sir Ian Hamilton, *The Soul and Body of an Army* (1921)

11 I love you — Titan lover, My own storm-days' Titan.
Isaac Rosenberg, 'Girl to Soldier on Leave', *Poems* (1922)

12 So we must say Goodbye, my darling, And go, as lovers go, for ever; Tonight remains, to pack and fix on labels And make an end of lying down together.
Alun Lewis, 'Goodbye', *Raider's Dawn* (1942)

13 When a subaltern applied for leave some

colonels would ask him how he intended to spend it, and if it appeared that he was merely going to poodlefake, they refused the leave. The young man must pursue animals, not girls.
John Masters, *Bugles and a Tiger* (1956)

301 LOYALTY

1 Young men, stand by each other and fight. Set no example of flight or fear, which bring shame, but make hearts in your breasts strong and courageous, taking no thought of your lives as you fight with men.
Tyrtaeus, address to the Spartan army before the Second Messenian War, 660 BC

2 The superior man is intelligently, not blindly, faithful.
Confucius, *Analects*, XV, 36 (*c.* 500 BC)

3 Loyalty is the the holiest good in the human heart.
Seneca, *Ad Lucilium*, Epistle 88

4 Thought shall be harder, heart the keener, courage the greater as our might lessens. Here lies our leader all hewn down the valiant man in the dust; may he lament for ever who thinks now to turn from this war play.
Anon., *The Battle of Maldon* (*c.*1000)

5 You may tell the King your father from me, that he may send us to the hottest spot in his dominions — to hell if he like — and I'll go at the head of them.
Lt-Gen Sir Alan Cameron of Erracht, letter to the Duke of York, 1796

6 The ties that bound us together were of the most sacred nature; they had been gotten in hardship and baptised in blood.
Theodore Gerrish, *Army Life: A Private's Reminiscence of the Civil War* (1882)

7 It is better to be faithful than famous.
Theodore Roosevelt, maxim adopted in 1903

8 Red lips are not so red
As the stained stones kissed by the English dead
Kindness of wooed and wooer
Seems shame to their love pure.
Wilfred Owen, 'Greater Love'

9 Loyalty is the marrow of honour.
Field Marshal Paul von Hindenburg, *Out of My Life* (1920)

10 Loyalty to petrified opinions never yet broke a chain or freed a human soul-
Mark Twain, 'Consistency', *Europe and Elsewhere* (1923)

11 People at home seem obsessed with the idea that the army will fight to the death to avenge Belgium. Nothing is further from the truth. We shall go on fighting until we are told to stop.
Captain J.E.H. Neville, *War Letters of a Light Infantryman* (1930)

12 Great achievements in war and peace can only result if officers and men form an indissoluble band of brothers.
Field Marshal Paul von Hindenburg, Proclamation to the German Army, 1934

13 There is a great deal of talk about loyalty from the bottom to the top. Loyalty from the top down is even more necessary and much less prevalent.
General George S. Patton, *War As I Knew It* (1947)

302 MACHINE GUNS

1 Onward Christian Soldiers, on to heathen lands,
Prayer-books in your pockets, rifles in your hands,
Take the glorious tidings where trade can be done:
Spread the peaceful gospel — with a Maxim gun.
Henri Labouchere, 'Pioneers' Hymn' (1872)

2 It occurred to me that if I could invent a machine — a gun — that would by its rapidity of fire enable one man to do as much battle duty as a hundred, that it would to a great extent, supersede the necessity of large armies, and consequently exposure to battle and disease would be greatly diminished.
Richard Gatling, inventor of the Gatling Gun, letter written in 1877

3 Whatever happens, we have got
The Maxim Gun, and they have not.
Hilaire Belloc, 'The Modern Traveller' (1898)

4 The machine gun man must be hot-blooded and dashing. He must have all the nerve and elan of the best light cavalry, all the resisting power of stolid and immovable infantry.
Lieutenant J.H. Parker, dispatch from the Spanish-American War, 1898

5 Machine guns play a great part — almost a decisive part under some conditions — in modern war.
Field Marshal Earl Haig, final dispatch after the Battle of the Somme, 1916

6 (Of the Battle of the Somme, 1916) To the south of the wood Germans could be seen, silhouetted against the sky-line, moving forward. I fired at them and watched them fall, chuckling with joy at the technical efficiency of the machine.
Lt-Col G.S. Hutchison, *Machine-guns: Their History and Employment* (1938)

7 Machine guns, opening out ahead, began to traverse methodically across our front, like flails of death crossing and recrossing as they sprayed the advancing line.
Norman Gladden, Diary, 1917, quoted in *Ypres 1917* (1967)

8 A few machine-guns tapped, spiteful and spasmodic.
Siegfried Sasson, *Memoirs of an Infantry Officer* (1930

9 The long burst of a machine gun does not kill a battalion; indeed, some men in a line will almost certainly pass through the run of bullets, in the gaps between them, so to speak. But if the line of men perseveres with a determined gallantry, over a long open approach, the end is certain.
General Anthony Farrar-Hockley, *The Somme* (1966)

10 (Of the Machine Gun Corps) No military pomp attended its birth or decease. It was not a great regiment with glamour and what not, but a great fighting corps, born for war only, and not for parades.
George Coppard, *With a Machine Gun to Cambrai* (1969)

11 (Of the Battle of Somme, 1916) The appearance of the machine-gun had not so much *disciplined* the act of killing — which was what seventeenth-century drill had done — as *mechanized* or *industrialized* it.
John Keegan, *The Face of Battle* (1976)

12 German guns cut our men down like

swathes in a cornfield, sounding as harmless as the seeng-seeng-seeng of a canary, a curious metallic chirping.
Denis Winter, *Death's Men* (1978)

---

303 MARCHING

---

1 Our life was but a battle and a march
And like the wind's blast, never-ending, homeless,
We stormed across the war-convulsed heath.
Friedrich von Schiller, *Wallensteins Tod*, III, 15 (1799), trs. Coleridge

2 March, march, Ettrick and Teviotdale,
Why the deil dinna ye march forward in order?
March, march, Eskdale and Liddesdale,
All the Blue Bonnets are bound for the Border.
Sir Walter Scott, 'Blue Bonnets Over the Border', *The Monastery* (1820)

3 The strength of an army, like power in mechanics, is reckoned by multiplying the mass by the rapidity; a rapid march increases the morale of an army, and increases its means of victory. Press on!
Napoleon I, *Maxims of War* (1831)

4 One noon day, at my window in the town,
I saw a sight — saddest that eyes can see —
Young soldiers marching lustily
Unto the wars,
With fifes, and flags in mottoed pageantry;
While all the porches, walks and doors
Were rich with ladies cheering royally.
Herman Melville, 'A Reverie: October 1861'

5 Hark! I hear the tramp of thousands, and of armed men the hum,
Lo! a nation's hosts have gathered, round one quick, alarming drum —
Saying, 'Come Freemen, come! Ere your heritage be wasted,' said the quick, alarming drum.
Francis Brett Harte, 'Reveille', *The Lost Galleon* (1867)

6 How good bad music and bad reason sound when one marches against an enemy!
Friedrich Nietzsche, *The Dawn* (1881)

7 Boots — boots — boots — boots — movin'
   up an' down again!
Rudyard Kipling, 'Boots', *Barrack-Room Ballads* (1892)

8 Far and near and low and louder
Dear to friends and food for powder,
Soldiers marching, all to die.
A.E. Housman, *A Shropshire Lad* (1896)

9 Oh, weren't they the fine boys! You never
   saw the beat of them,
  Singing all together with their throats
   bronze-bare;
  Fighting-fit and mirth-mad, music in the feet
   of them,
  Swinging on to glory and the wrath out
   there.
Robert Service, 'Tipperary Days', *Rhymes of a Red-Cross Man* (1916)

10 O Sing, marching men,
Till the valleys ring again
Charles Sorley, 'All the Hills and Vales Along', *Marlbrough and Other Poems* (1916)

11 Like flaming pendulums, hands
Swing across the khaki —
Mustard-coloured khaki —
To the automatic feet.
Isaac Rosenberg, 'Marching', *Poems* (1922)

12 They just stared straight ahead with eyes
that seemed to see nothing, and kept on
following the man in front — some in pain,
some asleep on their feet, some choked with
sickness, many limping — but all managing
to force one foot past the other in that
steady, subconscious, mechanical rhythm
which is the secret of the Infantry.
Fred Majdalany, *The Monastery* (1950)

---

304 MASSACRES

---

1 So ye shall not pollute the land wherein ye
are; for blood it defileth the land: and the
land cannot be cleansed of the blood that is
shed therein, but by the blood of him that
shed it.
Numbers, 35; 33

2 The opening barrage destroyed about six
thousand men on each side. Rifle-fire which
followed rid this best of worlds of about
nine or ten thousand villains who infested its
surface. Finally the bayonet provided
'sufficient reason' for the death of several
thousand more.
Voltaire, *Candide* (1759), trs. Butt

3 The soldier smiling hears the widow's cries;
And stabs the son before the mother's eyes.
With like remorse his brother of the trade,
The butcher, fells the lamb beneath the
   blade.
Jonathan Swift, 'On Dreams', *Miscellanies* (1727)

4 With fire and sword the country round
Was wasted far and wide,
And many a childing mother then,
And new-born baby died;
But things like that, you know, must be
At every famous victory.
Robert Southey, 'The Battle of Blenheim' (1798)

5 The tumult of each sacked and burning
   village;
The shout that every prayer for mercy
   drowns;
The soldiers' revels in the midst of pillage;
The wail of famine in beleaguered towns.
Henry Wadsworth Longfellow, 'The Arsenal at Springfield', *The Belfry at Bruges* (1843)

6 Towns without people, ten times took,
An' ten times left and burned at last;
An' starvin' dogs that come to look
For owners when a column passed.
Rudyard Kipling, 'The Return'

7 Have neither pity nor nerves — they are not
needed in war. Stamp out whatever pity and
compassion you may feel, and kill every
Russian, every Soviet person. Do not
hesitate — if there is an old man, a woman
or child before you — kill, and you will
thereby save yourself, ensure a future for
your family and cover yourself with eternal
glory.
Adolf Hitler, Order of the Day before the
invasion of the Soviet Union, 1941

8 There can be little doubt that in the combat
situation it becomes often meaningless to
ask the soldier to make fine discriminations
that distinguish a 'legitimate' act of war
from a war crime.
Peter Bourne, *Men, Stress and Vietnam* (1970)

9 We thought, we will go to Vietnam and be

Audie Murphys. Kick in the door, run in the hooch, give it a good burst — kill.
Lieutenant William Calley, evidence at the court-martial for the My Lai massacre, 1971

10 War always reaches the depths of horror because of idiots who perpetuate terror from generation to generation under the pretext of vengeance.
Guy Sajer, *The Forgotten Soldier* (1971)

305 MEDALS

1 It is not titles that honour men, but men that honour titles.
Niccolo Machiavelli, *Discorsi*, XXXVIII (1531)

2 Glory is the true and honourable recompense of gallant actions.
Alain René Le Sage, *Gil Blas* (1735)

3 Orders and decorations are necessary in order to dazzle the people.
Napoleon I, *Political Maxims* (1848)

4 It is not for pay and awards that the Prussian officer serves. An award or decoration flatters him but adds nothing to his merit in his own eyes or in those of his equals.
Prince Frederick Charles of Prussia, *The Origins and Development of the Spirit of the Prussian Officer* (1860)

5 The General got 'is decorations thick
(The men that backed 'is lies could not
        complain),
The Staff 'ad DSOs till we was sick,
 'An' the soldier — 'ad the work to do again!
Rudyard Kipling, 'Stellenbosch',
Barrack-Room Ballads of the Boer War (1902)

6 The number of medals on an officer's breast varies in inverse proportion to the square of the distance of his duties from the front lines.
C.E. Montague, *Fiery Particles* (1915)

7 I have my medals? — Discs to make my eyes
        close.
My glorious ribbons? — Ripped from my
        own back
In scarlet shreds. (That's for your poetry
        book.)
Wilfred Owen, 'A Terre', 1917

8 It is not in outward honours but in the inward satisfaction derived from duties done that a true soldier seeks his reward.
Colonel-General Hans von Seeckt, *The Principles of Army Training* (1921)

9 A ribbon is the only prize in war for the ordinary soldier.
Rowland Fielding, *War Letters to a Wife* (1929)

10 The safest thing to be said is that nobody knew how much a decoration was worth except the man who received it.
Siegfried Sassoon, *Memoirs of an Infantry Officer* (1930)

11 A medal glitters but it casts a shadow.
Winston S. Churchill, speech in House of Commons, 1944

12 Civilians may think it's a little juvenile to worry about ribbons, but a civilian has a house and a bankroll to show what he's done for the past four years.
William H. Mauldin, *Up Front* (1945)

13 It is a great thing for a unit to own a man who is awarded the Victoria Cross. The outside world thinks only of the individual; his unit regards the medal as partly theirs.
Patrick Davis, *A Child at Arms* (1970)

14 I am more proud of winning a Military Cross at Passchendaele than of any other achievement in my largely unsuccessful career.
Charles Carrington, quoted in Holmes, *Firing Line* (1985)

MILITARY HISTORY
See 335. WAR LITERATURE

306 MORALE

1 You know, I am sure, that not numbers or strength bring victory in war; but whichever army goes into battle stronger in soul, their enemies generally cannot withstand them.
Xenophon, *Anabasis*, III, i

2 There is a soul to an army as well as to the individual man and no general can accomplish the full work of his army unless

he commands the soul of his men as well as their bodies and legs.
General William T. Sherman, letter to General Ulysses S. Grant, 1864

3 Too great stress cannot be laid on developing good morale, a soldierly spirit, and a determination in all ranks to achieve success at all costs.
Lt-Gen Sir Lancelot Kiggell, Army Order, 1915

4 Fighting patrols are the finest stiffeners of morale.
Captain Edwin Campion Vaughan, Diary, 1917

5 War is a contest of nerves. That army wins which in times of tribulation can longest bear the mental strain of plague, pestilence and sudden death.
Charles Carrington, *A Subaltern's War* (1929)

6 Wars may be fought with weapons, but they are won by men. It is the spirit of men who follow and of the man who leads that gives the victory.
General George S. Patton, *Cavalry Journal* (1933)

7 It is not the number of soldiers, but their will to win which decides battles.
Lord Moran, *The Anatomy of Courage* (1945)

8 Very many factors go into the building up of sound morale in an army, but one of the greatest is that the men be fully employed at useful and interesting work.
Winston S. Churchill, *The Gathering Storm* (1948)

9 Loss of hope, rather than loss of life, is the factor that really decides wars, battles, and even the smallest combats.
Captain Sir Basil Liddell Hart, *Defence of the West* (1950)

10 The morale of the civilians in a city which has not yet been touched by war is seldom as high as it is among the soldiers in the frontline.
Alan Moorehead, *Gallipoli* (1956)

11 High morale means that every individual in a group will work — or fight — and, if needed, will give his last ounce of effort in its service.
Field Marshal Viscount Slim, *Courage and Other Broadcasts* (1957)

12 Battles are won primarily in the hearts of men.
Field Marshal Viscount Montgomery, *Memoirs* (1958)

13 The morale of the soldier is the greatest single factor in war.
Field Marshal Viscount Montgomery, *Memoirs* (1958)

MUNITIONS
See 269. AMMUNITION

307 MUSIC

1 He saith among the trumpets, Ha, ha; and he smelleth the battle afar off, the thunder of the captains, and the shouting.
Job 39:25

2 If the trumpet give an uncertain sound, who shall prepare himself to the battle?
I Corinthians 14:8

3 Sound all the lofty instruments of war
And by that music let us all embrace,
For heaven to earth, some of us never shall
A second time do such a courtesy.
William Shakespeare, *Henry IV,* Part 1, V,ii

4 Make all our trumpets speak; give them all breath,
Those clamorous harbingers of blood and death. (Macduff)
William Shakespeare, *Macbeth,* V.v.

5 Farewell the neighing steed and the shrill trump,
The spirit-stirring drum, the ear-piercing fife
The royal banner, and all quality,
Pride, pomp and circumstance of glorious war!
William Shakespeare, *Othello,* III,iii

6 The double double beat
Of the thundring drum
Cries, hark the foes come;
Charge, charge, 'tis too late to retreat.
John Dryden, 'A Song for St Cecilia's Day' (1687)

7 But when our Country's cause provokes to
arms,
How martial music every bosom warms!
Alexander Pope, 'Ode for Musick on St
Cecilia's Day' (*c.*1708)

8 See the conquering hero comes!
Sound the trumpets, beat the drums!
Thomas Morrell, libretto for Handel's
oratorio *Joshua* (1748)

9 The best sort of music is what it should be
— sacred; the next best, the military, has
fallen to the lot of the devil.
Samuel Taylor Coleridge, *Table Talk*
(1833)

10 The sound of the drum drives out thought;
for that very reason it is the most military of
instruments.
Joseph Joubert, *Pensées* (1842), trs.
Lyttelton

11 *Bang-whang-whang* goes the drum,
 tootle-te-tootle the fife;
No keeping one's haunches still: it's the
 greatest pleasure in life
Robert Browning, 'Up at a Villa — Down in
the City', *Men and Women* (1855)

12 Oh, the brave music of a *distant* Drum!
Edward Fitzgerald, *The Rubáiyát of Omar
Khayyám* (1859)

13 Louder, nearer, fierce as vengeance,
Sharp and shrill as swords at strife,
Came the wild Macgregor's clan-call,
Stinging all the air to life.
John Greenleaf Whittier, 'The Pipes at
Lucknow', *In War Time* (1860)

14 On the idle hill of summer,
Sleepy with the flow of streams,
Far I hear the steady drummer
Drumming like a noise in dreams.
A.E. Housman, *A Shropshire Lad* (1896)

15 They would march away in the dark, singing
to the beat of drums.
Siegfried Sassoon, *Memoirs of a
Fox-Hunting Man* (1928)

---

308 MUTINY

---

1 A motley crew of mutineering soldiers who
have murdered their officers, torn asunder

the ties of discipline, and not succeeded in
discovering a man in whom to bestow
supreme command are certainly the body
least likely to organise a serious and
protracted insurrection.
Karl Marx, *The First Indian War of
Independence* (1859)

2 The moment there is a sign of revolt, mutiny
or treachery, of which the symptoms are not
unusually a swollen head, and a tendency to
incivility, it is wise to hit the Oriental
straight between the eyes, and to keep
hitting him thus, till he appreciates exactly
what he is, and who is who.
Maj-Gen Sir George Younghusband, *Forty
Years a Soldier* (1920)

3 To me as a soldier, any strike among
disciplined men means mutiny.
Maj-Gen B.E.W. Childs, *Episodes and
Reflections* (1930)

4 Mutinies are suppressed in accordance with
laws of iron which are eternally the same.
Adolf Hitler, speech in the Reichstag, 1934

5 Mutiny and revolution are words which do
not enter the vocabulary of a German
soldier.
General Ludwig Beck, of the German
officers' plot to assassinate Hitler, in which
he took part, 1944, quoted in Reynolds,
Treason was no Crime (1976)

6 Where soldiers get into trouble of this
nature, it is nearly always the fault of some
officer who has failed in his duty.
Field Marshal Viscount Montgomery,
quoted in Ahrenfeldt, *Psychiatry in the
British Army in the Second World War*
(1958)

7 No one man can make eight obey him if
they are resolved to disobey.
Charles Carrington, *Soldier from the Wars
Returning* (1965)

8 Soldiers in fighting units are usually more
resistant to mutiny than those in depots and
training establishments.
Richard Holmes, *Firing Line* (1985)

9 Mutiny, armed or otherwise, has always
been the gravest of military crimes.
J.M. Brereton, *The British Soldier* (1986)

10 Mutiny, a premeditated, collective refusal to
obey orders, is the most heinous of crimes in
all armies, and though unthinkable, fear of it

is never far from the surface of
consciousness in the military mind.
Dominick Graham and Shelford Bidwell,
Tug of War: The Battle for Italy 1943–1945
(1986)

309 NO MAN'S LAND

1 This was a kind of Border, that might be
called No Man's Land, being a part of
Grand Tartary.
Daniel Defoe, *Robinson Crusoe* (1719)

2 No Man's Land was always the best part of
the front. There were no shells there, only,
perhaps, a Boche as frightened as oneself.
Major Desmond Allhausen, Diary, 1917,
quoted in Warner, *Passchendaele* (1987)

3 No-man's-land, much wider in places than I
had realised from any map, looked like a
long-neglected race-course by reason of the
distinctive greenness of its bare but relatively
undisturbed turf.
Billy Bishop, *Winged Warfare* (1918)

4 It was a place where a man of strong spirit
might know himself utterly powerless
against death and destruction, and yet stand
up and defy gross darkness and stupefying
shellfire, discovering in himself the
invincible resistance of an animal or an
insect, and an endurance which he might, in
other days, forget or disbelieve.
Siegfried Sassoon, *Memoirs of an Infantry
Officer* (1930)

5 To hear a wretch crying out in
no-man's-land and to be unable to help him
was enough to age anyone ten years
overnight.
Raleigh Trevelyan, *The Fortress* (1956)

6 This side of our wire everything is familiar
and every man a friend; over there, beyond
their wire, is the unknown, the enemy.
Charles Carrington, *Soldier from the Wars
Returning* (1965)

310 OATHS OF ALLEGIANCE

1 I'll take my corporal oath on it.
Miguel de Cervantes, *Don Quixote* (1605)

2 Oaths are wounds that a man stabs unto
himself.
Thomas Dekker, *The Seven Deadly Sins of
London,* II, 21 (1606)

3 My children, an oath is not an aria; one
can't simply give an encore.
Marquis de Lafayette, opposing demands to
make King Louis XVI re-take his oath
before the National Constituent Assembly,
1790

4 A soldier's vow to his country is that he will
die for the guardianship of her domestic
virtue, of her righteous laws, and of her
anyway challenged or endangered honour.
John Ruskin, *The Crown of Wild Olives*
(1866)

5 Well, we've both been lucky devils, both,
And there's no need of pledge or oath
To bind our lovely friendship fast,
By firmer stuff bound fast enough.
Robert Graves, 'Two Fusiliers', *Fairies and
Fusiliers* (1917)

6 Our calling is the defence of the Fatherland;
our country, therefore, the Army and every
soldier in it must feel the most intense
affection and be ready, in accordance with
his oath, to give his life itself in the
fulfilment of his duty.
Colonel-General Hans von Seeckt, *The
Principles of Army Training* (1921)

7 A soldier's honour requires him to pledge
the whole of himself for his people and the
Fatherland, even to the sacrifice of his life.
Field Marshal Paul von Hindenburg,
Proclamation to the German Army, 1934

8 For me, only half German, this ceremony
may have had even more significance than
for the others. Despite all the hardships we
had been through, my vanity was flattered
by my acceptace as a German amongst
Germans, and as a warrior worthy of
bearing arms.
Guy Sajer, *The Forgotten Soldier* (1971)

311 ORDERS

1 For I am a man under authority, having
soldiers under me: and I say to this man,
Go, and he goeth; and to another, Come,

and he cometh; and to my servant, Do this, and he doeth it.
Matthew 8:9

2 He who wishes to be obeyed must know how to command.
Niccolo Machiavelli, *Discorsi* (1531)

3 Who hath not served cannot command.
John Florio, *First Fruites* (1578)

4 There is great force hidden in a sweet command.
George Herbert, *Outlandish Proverbs* (1640)

5 Of all men Soldiers are most strictly tied to obedience, the want whereof may prove of very dangerous consequences.
R. Ram, *The Souldier's Catechism* (1644)

6 In time of peril, like the needle to the lodestone, obedience, irrespective of rank, generally flees to him who is best fitted to command.
Herman Melville, *Moby Dick* (1851)

7 Men must be habituated to obey or they cannot be controlled in battle, and the neglect of the least important order impairs the proper influence of the officer.
General Robert E. Lee, Circular to the Army of North Virginia, 1865

8 English soldiers are brought up with the idea that obedience is of more importance than initiative.
G.F.R. Henderson, *The Science of War* (1905)

9 The power to command has never meant the power to remain mysterious.
Marshal Ferdinand Foch, *Precepts* (1919)

10 In actions it is better to order than to ask.
General Sir Ian Hamilton, *Gallipoli Diary* (1920)

11 Soldiers who want to see whether an order suits their own ideas before they carry it out are absolutely worthless.
General Wilhelm Gröner, circular to German Army officers, 1930, quoted in Schüddekopf, *Das Heer und die Republik* (1955)

12 Operation orders do not win battles without the valour and endurance of the soldiers who carry them out.
Field Marshal Earl Wavell, *Soldiers and Soldiering* (1953)

PACIFISM
See 280. CONSCIENTIOUS OBJECTORS

312 PARADES

1 We spent the maist o' a' oor time
Jist marchin' up an' doon, sir,
Wi' a feathered bonnet on ma heid
An' poothered tae the croon, sir.
Anon., song of the 18th century, 'The Forfar Sodger,', quoted in Buchan and Hall, *The Scottish Folksinger* (1973)

2 It is indeed a noble and brilliant sight; to see the gallant defenders of their country drawn up in brilliant array before its peaceful citizens.
Charles Dickens, *Pickwick Papers* (1837)

3 Man is a military animal,
Glories in gunpowder, and loves parades.
Philip James Bailey, *Festus* (1839)

4 Marching along, fifty-score strong
Great-hearted gentlemen, singing this song.
Robert-Browning, 'Cavalier Tunes', *Bells and Pomegranates* (1841–6)

5 Whenever there are troops and leisure for it, there should be an attempt at military display.
Winston S. Churchill, memo to the Secretary of State for War, 1940

6 Young Fusiliers, strong-legged and bold,
March and wheel and march again.
Siegfried Sassoon, 'In Barracks', *Collected Poems* (1961)

7 This is our display of pride, our publicity, and we are ready to show what good soldiers look like.
Charles Carrington, *Soldier from the Wars Returning* (1965)

8 Military ritual is more than the delight of martinets, the bane of perennially scruffy soldiers and the abiding interest of a whole sub-species of military historians. It is a comprehensive framework of behaviour designed, *inter alia*, as a precaution against

disorder and a defence against the randomness of battle.
Richard Holmes, *Firing Line* (1985)

313 PAY

1 Stability between nations cannot be maintained without armies, nor armies without pay, nor pay without taxation.
Tacitus, *Histories*, IV, 74

2 In return for mead in the hall and drink of
        wine
He hurled his spears between two armies.
Aneirin, *The Gododdin* (c.600)

3 My desire is that if there be no pay like to come to me by the latter end of the week I may know it; I not being able to stay amongst them to hear the crying necessity of the hungry soldiers.
Earl of Essex, report to Parliament, 1643

4 If your pay and allowances for officers will not support them decently, then you will have only rich men who serve for pleasure or adventure, or inadequate wretches devoid of spirit.
Marshal Maurice de Saxe, *Mes Rêveries* (1757)

5 The soldier should not have any ready money. If he has a few coins in his pocket, he thinks himself too much of a lord to follow his profession, and he deserts at the opening of a campaign.
King Frederick II of Prussia, *Instructions for his Generals* (1747)

6 Always grumble and make difficulties when officers go to you for money that is due to them; when you are obliged to pay them endeavour to make it appear granting them a favour, and tell them they are lucky dogs to get it.
Francis Grose, *Advice to the Officers of the British Army* (1782)

7 How happy's the soldier who lives on his
        pay,
And spends half-a-crown out of sixpence a
        day.
John O'Keefe, 'The Poor Soldier' (1783)

8 For a soldier I listed, to grow great in frame,
And be shot at for sixpence a day.
Charles Dibdin, 'Charity', quoted in *The Professional Life of Mr Dibdin* (1803)

9 We are terribly distressed for *money*. I am convinced that £300,000 would not pay our debts, *and 2 months pay is due to the army.*
Duke of Wellington, letter to John Charles Villiers, 1809

10 Then though up late and early,
Our pay comes so rarely,
The divil a farthing we've ever to spare.
They say some disaster
Befell the Paymaster,
On my conscience I think that the money's
        not there!
Charles Lever, 'The Irish Dragoon' (1840)

11 Give him one farthing more than he really wants, and he gives way to his brutal propensities and immediately gets drunk.
Captain Henry Clifford, letter from the Crimean War, 1856

12 My saddest memory of the war is my continual state of poverty.
George Coppard, *With a Machine Gun to Cambrai* (1969)

314 PEACE

1 They shall beat their swords into plowshares, and their spears into pruninghooks: nation shall not lift up sword against nation, neither shall they learn war any more.
Isaiah 2:4

2 For peace, with justice and honour, is the fairest and most profitable of possession, but with disgrace and shameful cowardice it the most infamous and harmful of all.
Polybius, *Histories*, IV

3 A bad peace is even worse than war.
Tacitus, *Annals*, III

4 The most disadvantageous peace is better than the most just war.
Erasmus, *Adagia* (1508)

5 Peace hath her victories
No less renowned than war.
John Milton, 'To the Lord General Cromwell' (1652)

6 There never was a good war or a bad peace.
Benjamin Franklin, letter to Josiah Quincy,
1773

7 There is nothing so likely to produce peace
as to be well prepared to meet an enemy.
General George Washington, letter to
Elbridge Gerry, 1780

8 To be prepared for war is one of the most
effectual means of preserving peace.
General George Washington, speech to
Congress, 1790

9 An honourable peace is attainable only by
an efficient war.
Henry Clay, speech in House of
Representatives, 1813

10 Again and again we have owed peace to the
fact that we are prepared for war.
Theodore Roosevelt, speech to Naval War
College, Annapolis, 1897

11 War should never be entered upon until
every agency of peace has failed.
William McKinley, Inaugural Address,
1897

12 Peace, like war, can succeed only where
there is a will to enforce it, and when there is
available power to enforce it.
Franklin D. Roosevelt, speech to Foreign
Policy Association, New York, 1944

13 Everyone suddenly burst out singing.
Siegfried Sassoon, 'Everyone Sang',
Collected Poems (1947)

14 'Peace' is when nobody's shooting. A 'just
peace' is when our side gets what it wants.
William H. Mauldin, quoted in Bott, *Loose
Talk* (1980)

15 During peace, the Army's primary mission is
deterrence — being so well-trained,
equipped and led that no potential
adversary would mistake our nation's ability
and resolve to defend our interests.
John O. Marsh Jr., Confirmation
Statement, Washington, 1981

315 PRISONERS OF WAR

1 If any fall alive into the enemies' hands we
shall make them a present of him, and they
may do what they like with their prey.
Plato, *Republic,* V, 468

2 Those vanquished in war are held to belong
to the victor.
Aristotle, *Politics* 1225a, 6–7

3 When the war is concluded, I am definitely
of the opinion that all animosity should be
forgotten, and that all prisoners should be
released.
Duke of Wellington, letter to E.S. Waring,
1804

4 There is but one honourable mode of
becoming a prisoner of war. That is, by
being taken separately; by which is meant,
by being cut off, entirely, and when we can
no longer make use of our weapons. In this
case there can be no conditions, for honour
can impose none. We yield to irresistable
necessity.
Napoleon I, *Maxims of War* (1831)

5 When by the labour of my 'ands
I've helped to pack a transport tight
With prisoners for foreign lands,
I ain't transported with delight.
I know it's only just an' right,
But yet it somehow sickens me,
For I've learned at Waterval
The meanin' of captivity.
Rudyard Kipling, 'Half-Ballad of Waterval',
Barrack-Room Ballads of the Boer War
(1902)

6 To be a prisoner has always seemed to me
about the worst thing that could happen to
a man.
John Buchan, *Greenmantle* (1916)

7 A prisoner of war is a man who tries to kill
you and fails, and then asks you not to kill
him.
Winston S. Churchill, quoted in the
Observer, 1952

8 Escape is not only a technique but a
philosophy. The real escaper is more than a
man equipped with compass, maps, papers,
disguise and a plan. He has an inner
confidence, a serenity of spirit which makes
him a pilgrim.
Airey Neave, *They Have Their Exits* (1953)

9 (Of German prisoners of war in Normandy,
1944) They were past caring. The figures
were bowed with fatigue, although they had
nothing to carry but their ragged uniforms
and their weary, hopeless, battle-drugged
bodies.
R.M. Wingfield, *The Only Way out* (1955)

10 (Of Soviet prisoners of war, Eastern Front, 1941) Were those really human beings, these grey-brown figures, these shadows lurching towards us, stumbling and staggering, moving shapes at their last gasp, creatures which only some last flicker of the will to live enabled them to obey the order to march? All the misery in the world seemed to be concentrated here.
Benno Zieser, *In Their Shallow Graves* (1956)

11 (Of his imprisonment by the Japanese, 1942) In between being beaten up by the Japanese and having hates and having executions and things, we went on as if our lives were going to last for ever. This was a kind of school we were in.
Laurens van der Post, BBC television interview, 1968, quoted in Lewin, *Freedom's Battle*, III (1969)

12 Freedom was not our immediate want; even the desire for freedom is a luxury: what we needed first was a drink of water, and, after that, something to eat.
Robert Garioch, *Two Men and a Blanket* (1975)

---

316 PROMOTION

1 For promotion cometh neither from the east, nor from the west, nor from the south.
Psalms 75:6

2   Ambition,
The soldier's virtue.
William Shakespeare, *Antony and Cleopatra*, III, i

3 Who dies i' the wars more than his captain can,
Becomes his captain's captain.
William Shakespeare, *Antony and Cleopatra*, III, i

4 To take a soldier without ambition is to pull off his spurs.
Francis Bacon, *Essays* (1625)

5 I have seen some extremely good colonels become very bad generals.
Marshal Maurice de Saxe, *Mes Rêveries* (1757)

6 If you ever wish to rise a step above your present degree, you must learn that maxim of the art of war, of currying favour with your superiors; and you must not only cringe to the commander-in-chief himself, but you must take special care to keep in with his favourites, and dance attendance on his secretary.
Francis Grose, *Advice to the Officers of the British Army* (1782)

7 It would be desirable that the only claim to promotion should be military merit; but this is a degree of perfection to which the disposal of military patronage has never been, and cannot be, I believe, brought in any military establishment.
Duke of Wellington, letter to Lt-Col Henry Torrens, 1810

8 More like a mother she were —
Showed me the way to promotion an' pay,
An' I learned about women from 'er!
Rudyard Kipling, 'The Ladies', *Barrack-Room Ballads* (1892)

9 Smooth answers smooth the path to promotion.
Captain Sir Basil Liddell Hart, *Thoughts on War* (1944)

10 An extensive use of weedkiller is needed in the *senior* ranks after a war; this will enable the first class young officers who have emerged during the war to be moved up.
Field Marshal Viscount Montgomery, *Memoirs* (1958)

11 By good fortune in the game of military snakes and ladders, I found myself a general.
Field Marshal Viscount Slim, *Unofficial History* (1959)

12 A military man lacking ambition has no chance to advance.
Martin Blumenson, *Mark Clark* (1984)

---

317 PUNISHMENT

1 Punishment and fear thereof are necessary to keep soldiers in order in quarters: but in the field they are more influenced by hope and rewards.
Vegetius, *De Re Militari* (378)

2 Frae a' bridewell cages and blackholes,
   And officers' canes, wi' their halbert poles,
   And frae the nine-tailed cat that opposes our
       souls
   Gude Lord deliver us.
   Sergeant Bauldy Corson, ballad of the late
   18th century, 'The Soldier's Prayer'

3 You may try an officer for surrendering up a
   fort when under no necessity to do it, but let
   not the blood of the poor be spilt profusely.
   Duke of Cumberland, remark, 1745

4 Soldiers of the present times have nothing
   but their bodies and can only be punished
   corporally.
   Francis Grose, *Advice to the Officers of the
   British Army* (1782)

5 Hall was sentenced to receive five hundred
   lashes for house-breaking; he got four
   hundred of them before he was taken down;
   and in the space of six weeks was judged
   able to sustain the remainder of his
   punishment, as his back was skinned over.
   Robert Hamilton, *The Duties of a
   Regimental Surgeon Considered* (1787)

6 Philanthropists who decry the lash ought to
   consider in what manner the good men —
   the deserving, exemplary soldiers — are to
   be protected.
   Quartermaster-Sergeant James Anton,
   1795, quoted in Glover, *Peninsular
   Preparation* (1963)

7 I never was any more afraid of the lash than
   I was of the gibbet, no man ever comes to
   that but through his own conduct.
   John Stevenson, quoted in Glover,
   Peninsular Preparation (1963)

8 Where soldiers are to be ruled, there is more
   logic in nine tails of a cat than in the mouth
   of a hundred orators.
   John Kincaid, *Random Shots from a
   Rifleman* (1847)

9 You are her Majesty's soldiers now,
   And if you care to wrangle,
   The cat-o'-nine-tails is your doom,
   Tied up to the triangle.
   Anon., song of 1835 'Calling out the Militia
   for Duty'

10 I detest the sight of the lash; but I am
   convinced that the British army can never go
   on without it.
   Rifleman John Harris, *Recollections* (1848)

11 The barbarity of the English military code
   incited public horror.
   Maj-Gen Sir William Napier, *History of the
   Peninsular War*, XXI (1851)

12 So it's pack-drill for me and a fortnight's CB
   For 'drunk and resisting the Guard'.
   Rudyard Kipling, 'Cells', *Barrack-Room
   Ballads* (1892)

13 In order to keep the honour of his unit
   bright, a commander may have to use his
   sword as a weapon of punishment,
   exceedingly shameful though it is to have to
   shed the blood of one's own soldiers on the
   battlefield.
   Maj-Gen Tanaka, Order of the Day before
   the Battle of Imphal, 1944

---

318 RATIONS

1 Soldiers' stomachs always serve them well.
   William Shakespeare, *Henry VI*, Part 1, II,
   iii

2 Have a keen eye in all ends and places of our
   duchy that the beer is not overcharged, also
   that bread and other victuals and all things
   else that man cannot forgo in his
   undertakings are brought and kept to a
   cheap rate.
   Albert von Wallenstein, letter to the
   Emperor Ferdinand II, 1627

3 An army cannot preserve good order unless
   its soldiers have meat in their bellies, coats
   on their backs and shoes on their feet.
   Duke of Marlborough, letter to Colonel
   William Cadogan, 1703

4 They'll fight all the better on empty bellies.
   Remember what a dessert they got to their
   dinner at Falkirk.
   Duke of Cumberland, Order before the
   Battle of Culloden, 1746

5 An army, like a serpent, marches upon its
   belly.
   King Frederick II of Prussia, attr.

6 The men, together with their officers, are
   like young ravens — they only know how to
   open their mouths to be fed.
   A.L.F. Schaumann, of the British Army in
   the Peninsular War, 1808, quoted in *On the
   Road with Wellington* (1924)

7 An army marches on its stomach.
   Napoleon I, attr.

8 All men require two pounds of *food* a day.
   Vegetable food is less convenient than
   Animal food, the last walking with you.
   Duke of Wellington, quoted in Rogers,
   Recollections (1859)

9 You talk o' better food for us, an' schools,
                        an' fires, an' all:
   We'll wait for extra rations if you treat us
                        rational.
   Don't mess about the cook-room slops, but
                        prove it to our face
   The Widow's Uniform is not the
                        soldier-man's disgrace.
   Rudyard Kipling, 'Tommy', *Barrack-Room
   Ballads* (1892)

10 Lack of food constitutes the single biggest
   assault upon morale.
   Bernard Fergusson, *The Wild Green Earth*
   (1946)

11 Tea was brewed three or four times a day,
   and if the supplies had ever failed the morale
   of the army would have been reduced more
   than by a major defeat.
   Roy Farran, *Winged Dagger* (1948)

12 The bully makes me bloody wild,
   I'd nearly eat a bloody child,
   The salty water makes me riled.
   Oh bloody! Bloody! Bloody!
   Hugh Patterson, 'Tobruk', quoted in Laffin,
   Digger (1959)

---

319 RELIGION

1 The enemies of God were blind and
   stupefied: their eyes could plainly see the
   knights of Christ, but it was as if they saw
   nothing, and they no longer dared to rise
   against the Christians, for the divine power
   terrified them.
   Nonymi Gesta Francorum (1101)

2 O God of battles! steel my soldiers' hearts.
   William Shakespeare, *Henry V*, IV, i

3 O Lord! thou knowest how busy I must be
   this day. If I forget thee do not forget me.
   Sir Jacob Astley, prayer for his men before
   the Battle of Edgehill, 1643

4 God calls Himself a Man of War, and Lord
   of Hosts; Abraham had a regiment of three
   hundred and eighteen trained men; David
   was employed in fighting the Lord's battles.
   R. Ram, *The Souldier's Catechism* (1644)

5 They that fight against the Church's
   enemies are God's helpers against the
   mighty. They are the instruments of justice
   and the executioners of God's judgements.
   R. Ram, *The Souldier's Catechism* (1644)

6 The soldier at the same time may shoot out
   his prayer to God, and aim his pistol at his
   enemy, the one better hitting the mark for
   the other.
   Thomas Fuller, *Good Thoughts for Bad
   Times* (1645)

7 It is said that God is not on the side of the
   heaviest battalions, but of the best shots.
   Voltaire, letter to M. le Riche, 1770

8 There is a time to pray and a time to fight.
   This is the time to fight.
   John Peter Gabriel Mühlenberg, sermon at
   Woodstock, Virginia, 1775

9 The chaplain is a character of no small
   importance in a regiment, though many
   gentlemen in the army think otherwise.
   Francis Grose, *Advice to the Officers of the
   British Army* (1782)

10 A soldier's life of honour is subject to so
   many changes that he has not time to think
   of religion like another man, for no sooner,
   perhaps, does he think of prayer than the
   drum beats or trumpets sound to arms —
   and how can he talk of forgiving his enemy
   when it is his whole duty to destroy them.
   Benjamin Miller, after the Battle of the Nile,
   1801, quoted in the *Journal of the Society
   for Army Historical Research* (1928)

11 O Lord, if Thou wilt not be for us today, we
   ask that Thou be not against us. Just leave it
   between the French and ourselves.
   Lt-Gen Sir Alan Campbell of Erracht,
   prayer before battle, *c.*1809

12 The Almighty God would never will it that a
   Christian army should be cut up by a pagan
   host. Halt, sir, or as a minister of the word
   of God, I'll shoot you.
   Rev. William Whiting, Order to the 14th
   Light Dragoons during the Battle of
   Chilianwala, 1849

13 Fight the good fight

With all thy might;
Christ is thy strength,
and Christ thy right.
John Monsell, 'Fight the Good Fight'
(*c.*1855)

14 Onward! Christian soldiers,
Marching as to war,
With the Cross of Jesus
Going on before.
Sabine Baring-Gould, 'Onward! Christian
Soldiers' (*c.*1870)

15 God heard the embattled nations sing and
      shout
'Gott strafe England!' and 'God save the
      King!'
God this, God that, and God the other
      thing —
'Good God!' said God, 'I've got my work
      cut out.'
J.C. Squire, *The Survival of the Fittest*
(1916)

16 I had a glimpse of my army chaplain now
and then, but never anywhere near the
trenches.
George Coppard, *With a Machine Gun to
Cambrai* (1969)

---

320 RIFLES

1 Firearms and not cold steel now decide
battles.
Marshal de Puységur, *Art de la Guerre*
(1748)

2 Teach the soldier how to load his musket
properly, whether with cartridge or loose
powder, with or without the bayonet fixed;
how he should give fire under the different
circumstances he will encounter; teach him
never to fire without an order, and never to
do so without aiming, thus avoiding the
waste of fire to no purpose.
Marshal de Puységur, *Art de la Guerre*
(1748)

3 Ready, be ready to meet the storm!
Rifleman, rifleman, rifleman form!
Lord Tennyson, 'Rifleman Form', *The
Times*, 1859

4 The outward and visible sign of the end of
war was the introduction of the magazine
rifle.
I.S. Bloch, *Is War Impossible?* (1899)

5 Shooting at a fixed target is only a step
towards shooting at a moving one, like a
man.
Lord Baden-Powell, *Scouting for Boys*
(1908)

6 The Golden Age is to come with the ability
of every ploughboy to handle a rifle.
James Anson Farrar, *Invasion and
Conscription* (1909)

7 Only the stuttering rifles' rapid rattle
Can patter out their hasty orisons.
Wilfred Owen, 'Anthem for Doomed
Youth' (1917)

8 The smallest unit is the single man with his
rifle.
Slogan adopted by the Palmach, the striking
force of the Haganah, 1941, quoted in
Allon, *The Making of Israel's Army* (1970)

9 The distance at which all shooting weapons
take effect screens the killer against the
stimulus sensation which would otherwise
activate his killing inhibitions.
Konrad Lorenz, *On Aggression* (1966)

10 The Lee Enfield was much valued as a
symbol of security, an assurance that a
man's fate was to a degree in his own hands.
Denis Winter, *Death's Men* (1978)

---

321 SANDHURST

1 Oh, what will the girls of Camberley say
If the RMC was to march away
And leave them all in the family way
With the hell of a doctor's bill to pay?
Anon., song of *c.*1910, sung by Sandhurst
officer cadets, quoted in Thomas, *The Story
of Sandhurst* (1961)

2 (Of the College's Governor, Lt-Gen Sir C.J.
East, 1897) What he governed I have no
idea, for the cadets saw him but twice a
term, when they arrived and when they
departed.
Maj-Gen J.F.C. Fuller, *Memoirs of an
Unconventional Soldier* (1936)

3 It was here at Sandhurst that the legend of
Haig as a man who must one day rise to the
highest command, originated.
Duff Cooper, *Haig* (1939)

4 This Academy trains future regular officers

to command and lead the finest soldiers in the world.
Maj-Gen F.R.G. Matthews, address to Sandhurst officer cadets, 1947

5 It was permissable, indeed almost laudable, to cheat if that was your only hope of passing out and getting your commission. What did book learning matter, anyway, to a subaltern of a fighting regiment?
John Masters, *Bugles and a Tiger* (1957)

6 The great tradition of which you are the heirs was built by those who served and studied her before you, upon the firm foundations of unselfishness, heroism and self-sacrifice.
Queen Elizabeth II, address to Sandhurst officer cadets, 1957

7 The competition for the infantry was keener, as life in the cavalry was so much more expensive. Those who were at the bottom of the list were accordingly offered the easier entry into the cavalry.
Winston S. Churchill, *My Early Life* (1959)

8 I found the modern Sandhurst an inspiring place to visit, with its stimulating blend of tradition and progress.
Brigadier Sir John Smyth, *Sandhurst* (1961)

9 In spite of my mishaps at Sandhurst, I have always had an affection for the place.
Field Marshal Viscount Montgomery, Preface to Smyth, *Sandhurst* (1961)

10 The grounds of the Royal Military Academy looked very unmilitary at first glance: more Capability Brown than Prussian.
Edmund Ions, *A Call to Arms* (1972)

---

322 SEX

1 Women adore a martial man.
William Wycherley, *The Plain Dealer,* II (1677)

2 And when we have done with mortars and
            guns:
'If you please, Madam Abbess, a word with
            your nuns.'
Each soldier shall enter the convent in buff,
And then, never fear, we will given them hot
            stuff.
Ned Botwood, song of 1774, 'Hot Stuff'

3 (Of the fall of Rome to the Goths) The

brutal soldiers satisfied their sensual appetites without consulting either the inclination or the duties of their female captives.
Edward Gibbon, *The Decline and Fall of the Roman Empire* (1776–88)

4 (Of the fall of Badajoz) I was told that very few females, old or young, escaped violation by our brutal soldiery, mad with brandy and with passion.
Sir James McGrigor, *Autobiography and Services* (1861)

5 I have told him that he may stay there forty-eight hours which is as long as any reasonable man can wish to stay in bed with the same woman.
Duke of Wellington, on dealing with an officer's request for leave during the Peninsular War, quoted in Guedalla, *The Duke* (1940)

6 The lieutenant-captain was superstitious and considered it a great sin to amuse himself with women before going into action.
Count Leo Tolstoy, *Tales of Army Life* (1855)

7 When the military man approaches, the world locks up its spoons and packs off its womankind.
George Bernard Shaw, *Major Barbara,* III (1907)

8 While treating all women with perfect courtesy, you should avoid any intimacy.
Field Marshal Earl Kitchener, Special Army Order, 1914

9 Après la guerre fini,
Tous les soldats partis,
Mademoiselle avec piccanini
Souvenir des Anglais.
Anon., song of the First World War

10 The bird of war is not the eagle, but the stork.
Charles Frances Potter, speech in the Senate, 1931

11 There's no doubt that German girls are bending over backwards to please Canadian soldiers.
Kate Aitken, radio broadcast, 1947, quoted in *Making Your Living is Fun* (1959)

12 It is as a soldier that you make love, and as a lover that you make war.
Antoine de Saint-Exupéry, *The Wisdom of the Sands* (1948), trs. Gilbert

13 A good soldier has his heart and soul in it.
When he receives an order, he gets a
hard-on, and when he sends his lance into
the enemy's guts, he comes.
Bertholt Brecht, *The Caucasian Chalk Circle*
(1949), trs. Bentley and Apelman

14 There is frequently a homosexual bond
between the leaders and the led.
Shelford Bidwell, *Modern Warfare* (1973)

323 SPORT

1 Most sorts of diversion in men, children and
other animals are an imitation of fighting.
Jonathan Swift, *Thoughts on Various
Subjects* (1711)

2 Being a good sportsman, a good cricketer,
good at rackets or any other manly game is
no mean recommendation for staff
employment.
Field Marshal Viscount Wolseley, *The
Soldier's Pocket-Book for Field Service*
(1869)

3 And it's not for the sake of a ribboned coat,
Or the selfish hope of a season's fame,
But his captain's hand on his shoulder
                    smote —
'Play up! play up! and play the game!'
Sir Henry Newbolt, 'Vitaï Lampada' (1897)

4 Our gallant fellows at the front are carrying
their football training into practice on the
battlefield. They are 'playing the game' in all
conscience.
Lord Baden-Powell, *Headquarters Gazette*
(1914)

5 War is the only sport that is genuinely
amusing. And it is the only sport that has
any intelligible use.
H.L. Mencken, *Prejudices* (1919)

6 It seemed something to be grateful for —
that the War hadn't killed cricket yet, and
already it was a relief to be in flannels and
out of uniform.
Siegfried Sassoon, *Memoirs of a
Fox-Hunting Man* (1928)

7 Serious sport has nothing to do with fair
play. It is bound up with hatred, jealousy,
boastfulness, disregard of all rules and

sadistic pleasure in witnessing violence: in
other words it is war minus the shooting.
George Orwell, 'The Sporting Spirit',
Shooting in Elephant (1950)

8 These plains were their cricket pitch
and in the mountains the tremendous drop
                    fences
brought down some of the runners. Here
                    then
under the stones and earth they dispose
                    themselves,
I think with their famous unconcern.
It is not gunfire I hear but a hunting horn.
Keith Douglas, 'Aristocrats', *Collected
Poems* (1951)

9 Pro football is like nuclear warfare, There
are no winners, only survivors.
Frank Gifford, *Sports Illustrated*, 1960

10 The team spirit inherent in all international
sport gives scope to a number of truly
valuable patterns of social behaviour which
are essentially motivated by aggression, and
which, in all probability, have evolved under
the selection pressure of tribal warfare at the
very dawn of culture.
Konrad Lorenz, *On Aggression* (1966)

11 Hunting was a substitute for war, both for
those taking part in it and those for whom it
was designed to frighten.
Bamber Gascoigne, *The Great Moguls*
(1971)

12 Sport concerned the military in two ways:
firstly, as the straight road to physical health
and strength, indispensable to the good
soldier; secondly, because of the special
value attributed to team games in training
the essential qualities of the officer and
leader.
Geoffrey Best, quoted in Simon and
Bradley, *The Victorian Public School* (1975)

324 STRATEGY

1 The theory of warfare tries to discover how
we may gain a preponderance of physical
forces and material advantages at the
decisive point. As this is not always possible,
theory also teaches us to calculate moral
factors.
Karl von Clausewitz, *Principles of War*
(1812)

2 Unhappy the general who comes on the field of battle with a system.
Napoleon I, *Maxims of War* (1831)

3 He who writes on strategy and tactics should force himself to teach an exclusive national strategy and tactics — which are the only ones liable to benefit the nation for whom he is writing.
Colmar von der Goltz, *The Nation in Arms* (1883)

4 What is grand strategy? Common-sense applied to the art of war.
General William T. Sherman, speech at Portland, Maine, 1890

5 No proposition Euclid wrote
No formula the text-books know
Will turn the bullet from your coat
Or ward the tulwar's downward blow.
Rudyard Kipling, 'Arithmetic on the Frontier', *Barrack-Room Ballads* (1892)

6 It is of paramount importance in war that there should be a definite plan of operation and that that plan should be carried out with promptness and decision.
Field Marshal Sir William Robertson, letter to Field Marshal Earl Kitchener, 1915

7 The civilian is too inclined to think that war is only the working out of an arithmetical problem with given numbers. It is anything but that. On both sides it is a case of wrestling with powerful, unknown physical and psychological forces, a struggle which inferiority in numbers makes all the more difficult.
General Erich von Ludendorff, *My War Memoirs* (1920)

8 In strategy the longest way round is apt to be the shortest way home!
Captain Sir Basil Liddell Hart, *The Way to Win Wars* (1942)

9 Grand strategy must always remember that peace follows war.
Captain Sir Basil Liddell Hart, *Thoughts on War* (1944)

10 The modern army commander must free himself from routine methods and show a comprehensive grip of technical matters, for he must be in a position continually to adapt his ideas of warfare to the facts and possibilities of the moment.
Field Marshall Erwin Rommel, quoted in Liddell Hart, *The Rommel Papers* (1953)

11 The theory of war and strategy is the core of all things.
Mao Tse-Tung, *Problems of War and Strategy* (1954)

12 Against a brave and well-led enemy all war hinges on the ability to destroy the enemy's main armed forces in pitched battle, or, of course, to subvert his will to fight at all.
John Masters, *The Road Past Mandalay* (1961)

325 SWEARING

1   Then a soldier,
Full of strange oaths, and bearded like the
                    pard.
William Shakespeare, *As You Like It*, II,vii

2 That in the captain's but a choleric word
Which in the soldier is flat blasphemy.
William Shakespeare, *Measure for Measure*, II, ii

3 This would make a saint swear like a trooper.
Sir Francis Beaumont and John Fletcher, *Philaster*, IV, 2 (1608)

4 'Our armies swore terribly in Flanders,' cried my uncle Toby, 'but nothing to this.'
Laurence Sterne, *Tristram Shandy*, vol. III (1759–67)

5 That unmeaning and abominable custom, swearing.
General George Washington, General Order to the Army, 1776

6 Though you are not to allow swearing in others, it being forbidden by the articles of war, yet by introducing a few oaths occasionally into your discourse, you will give your inferiors some idea of your courage; especially if you should be advanced in years: for then they must think you a dare-devil indeed.
Francis Grose, *Advice to the Officers of the British Army* (1782)

7 When angry, count four; when very angry, swear.
Mark Twain, *Pudd'nhead Wilson* (1894)

8 Th' best thing about a little judicyous swearin' is that it keeps the temper. 'Twas

intinded as a compromise between runnin'
and fightin'.
Finley Peter Dunne, *Observations by Mr
Dooley* (1902)

9   Most colonels can swear but there are
specialists.   .
Robert Blatchford, *My Life in the Army*
(1910)

10 The most common word in the mouths of
American soldiers has been a vulgar
expression for sexual intercourse. This word
does duty as adjective, adverb, verb, noun
and in any other form it can possibly be
used, however inappropriate or ridiculous in
application.
J. Glenn Gray, *The Warriors: Reflections on
Men in Battle* (1970)

326 SWORDS

1   The blade itself incites to violence.
Homer, *Odyssey,* XVI

2   His sword echoed in the heads of mothers.
Aneirin, *The Gododdin* (*c.*600)

3   (Of the Battle of Lake Pskov, 1242) There
was great slaughter and the clash of spears
shivering and the clang of swords hewing as
they strove on the frozen sea: the ice could
not be seen; it was covered with blood.
Novgorod Chronicle (1242)

4   Our swords shall play the orators for us.
Christopher Marlowe, *Tamburlaine the
Great,* Part 1, I, ii

5   Full bravely has thou fleshed
Thy maiden sword.
William Shakespeare, *Henry IV,* Part 1, V,
iv

6   (Of the Battle of Prestonpans, 1745) The
field of battle presented a spectacle of
horror, being covered with hands, legs and
arms, and mutilated bodies; for the killed all
fell by the sword.
Chevalier James Johnstone, *Memoirs*
(1820)

7   Step by step, in the past, man has ascended
by means of the sword.
Rear-Admiral A.T. Mahan, *The Peace
Conference* (1899)

8   He knew men and named me
The War-Thing, the Comrade,
Father of honour, and giver of kinship,
The fame-smith, the song-master,
Bringer of women.
W.E. Henley, 'The Song of the Sword', *For
England's Sake* (1900)

9   The first dry rattle of new-drawn steel
Changes the world today!
Rudyard Kipling, 'Before Edgehill',
Fletcher, *A History of England* (1911)

10 The sword stroke, practised a thousand
times, polished and refined and measured to
pass unerringly beneath an opponent's
parry, was beaten flat by the musket-shot.
John Keegan, *The Face of Battle* (1976)

327 TACTICS

1   It is an invariable axiom of war to secure
your own flanks and rear and endeavour to
turn those of your enemy.
King Frederick II of Prussia, *Instructions for
his Generals* (1747)

2   It is an accepted maxim of war, never to do
what your enemy wishes you to do, for this
reason alone, that he desires it.
Napoleon I, *Maxims of War* (1831)

3   Tactics is an art based on the knowledge of
how to make men fight with maximum
energy against fear, a maximum which
organisation alone can give.
Charles Ardant du Picq, *Battle Studies*
(1870)

4   I think if instead of Minor Tactics or books
on the art of war we were to make our
young officers study Plutarch's *Lives* it
would be better; there we see men
(unsupported by any true belief — pure
pagans) making as a matter of course their
lives a sacrifice, but in our days it is the
highest merit not to run away.
General Charles George Gordon,
Khartoum Journal, 1884

5   In the stage of the wearing out struggle
losses will necessarily be heavy on both
sides, for in it the price of victory is paid.
Field Marshal Earl Haig, *Dispatches* (1919)

6   Nine-tenths of tactics are certain, and taught

in books: but the irrational tenth is like the kingfisher flashing across the pool and that is the test of generals. It can only be ensured by instinct, sharpened by thought, practising the stroke so often that at the crisis it is as natural as reflex.
T.E. Lawrence, 'The Science of Guerrilla Warfare', Encyclopaedia Britannica (1929)

7   I rate the skilful tactician above the skilful strategist, especially him who plays the bad cards well.
Field Marshal Earl Wavell, *Soldiers and Soldiering* (1939)

8   In war, the only sure defense is offense, and the efficiency of offense depends on the warlike souls of those conducting it.
General George S. Patton, *War As I Knew It* (1947)

9   There is no approved solution to any tactical situation.
General George S. Patton, *War As I Knew It* (1947)

10 The commander must decide how he will fight the battle *before it begins*. He must then decide how he will use the military effort at his disposal to force the battle to swing the way he wishes it to go; he must make the enemy dance to his tune from the beginning, and never vice versa.
Field Marshal Viscount Montgomery, *Memoirs* (1958)

328 TANKS

1   I'd like to see a Tank come down the stalls,
Lurching to rag-time tunes, or 'Home, sweet Home',
And there'd be no more jokes in Music-halls
To mock the riddled corpses round Bapaume.
Siegfried Sassoon, 'Blighters', *Counter-Attack* (1918)

2   It had been instilled into us that a tank commander's duty was to stick to his tank to the end, his whole object being to protect it. If need be, he must, like the captain of a ship, perish with it.
Captain D.E. Hickey, *Rolling into Battle* (1920)

3   The tank marks as great a revolution in land

warfare as an armoured steamship would have marked had it appeared amongst the toilsome triremes of Actium.
Colonel-General Heinz Guderian,7 Achtung! Panzer! (1937)

4   The tank was the beginning of the bullet-proof army.
Winston S. Churchill, *The World Crisis*, II (1923)

5   From a mockery the tanks have become a terrible weapon. Armoured they come rolling on in long lines, more than anything else embodying for us the horror of war.
Erich Maria Remarque, *All Quiet on the Western Front* (1929)

6   Wherever in future wars the battle is fought, tank troops will play the decisive role.
Colonel-General Heinz Guderian, *Achtung! Panzer!* (1937)

7   With the development of tank forces the old linear warfare is replaced by circular warfare.
Captain Sir Basil Liddell Hart, *Thoughts on War* (1944)

8   The officers of a panzer division must learn to think and act independently within the framework of the general plan and not wait until they receive orders.
Field Marshal Erwin Rommel, quoted in Liddell Hart, *The Rommel Papers* (1953)

9   The whole thing was unlike a boxing match as it could be, because in a tank battle the first hit was the winning one.
Peter Elstob, *Warriors for the Working Day* (1960)

10 Tanks burn in a way that has its own grotesque poignancy.
David Holbrook, *Flesh Wounds* (1966)

11 We felt a sense of fun,
Ridiculous, half-sorrowful for horse,
Canned soldiers, with no proper war to fight,
Living for the day.
George MacBeth, *A War Quartet* (1969)

12 They are an elite, men of high spirit, like submarine men too — their comradeship forged by shared hazards and the shared intoxication of manning intricate, almost invincible machines.
David Irving, *The Trail of the Fox: The Life of Field-Marshal Erwin Rommel* (1977)

13 It takes twenty minutes for a medium tank to incinerate; and the flames burn slowly; so figure it takes ten minutes for a hearty man within to perish. You wouldn't even be able to struggle for chances are both exits would be sheeted with flame and smoke. You would sit, read *Good Housekeeping*, and die like a dog. Steel coffins indeed!
Nat Frankel, *Patton's Best* (1978)

329 TRAINING
See also 288. DRILL, 287. DISCIPLINE, 291.
ESPRIT DE CORPS

1 To lead an untrained people to war is to throw them away.
Confucius, *Analects* XIII (*c*.500 BC)

2 We must remember that one man is much the same as another, and that he is best who is trained in the severest school.
Thucydides, *Peloponnesian Wars*

3 The courage of a soldier is heightened by his knowledge of his profession.
Vegetius, *De Re Militari* (378)

4 To bring Men to a proper degree of Subordination, is not the work of a day, a month, or even a year.
General George Washington, letter to the President of Congress, 1776

5 Hardship, poverty and want are the best school for a soldier.
Napoleon I, *Maxims of War* (1831)

6 Body and spirit I surrendered whole
To harsh instructors — and received a soul.
Rudyard Kipling, 'The Wonder', *Epitaphs of the War* (1919)

7 No study is possible on the battlefield; one does there simply what one can in order to apply what one knows. Therefore, in order to do even a little, one has already to know a great deal and know it well.
Marshall Ferdinand Foch, *Precepts* (1919)

8 At first astonished, then embittered, and finally indifferent, we recognised that what matters is not the mind but the boot brush, not intelligence but the system, not freedom but drill.
Erich Maria Remarque, *All Quiet on the Western Front* (1929)

9 In no other professions are the penalties for employing untrained personnel so appalling or irrevocable as in the military.
General Douglas MacArthur, Annual Report to the Chiefs of Staff, Army, 1933

10 War makes extremely heavy demands on the soldiers' strength and nerves. For this reason make heavy demands on your men in practice.
Field Marshal Erwin Rommel, *Infantry Attacks* (1937)

11 The best form of 'welfare' for the troops is first-class training.
Field Marshal Erwin Rommel, quoted in Liddell Hart, *The Rommel Papers* (1953)

12 To-day we have naming of parts. Yesterday
We had daily cleaning. And tomorrow morning,
We shall have what to do after firing. But to-day,
To-day we have naming of parts.
Henry Reed, 'Naming of Parts', *A Map of Verona* (1946)

13 Long training tends to make a man more expert in execution, but such expertness is apt to be gained at the expense of fertility of ideas, originality and elasticity.
Captain Sir Basil Liddell Hart, *Defence of the West* (1950)

14 One advantage of exceptionally hard training is that it proves to a man what he can do and suffer.
Brigadier Michael Calvert, *Prisoners of Hope* (1971)

330 TRENCHES

1 When I hear talk of lines, I always think I am hearing talk of the walls of China. The good ones are those that nature has made, and the good entrenchments are good dispositions and brave soldiers.
Marshal Maurice de Saxe, *Mes Rêveries* (1757)

2 Everybody will be entrenched in the next war. It will be a great war of entrenchments. The spade will be as indispensible to a soldier as his rifle.
I.S. Bloch, *Is War Impossible?* (1899)

3 The firing-trench is our place of business —

our office in the city, so to speak. The
supporting trench is our suburban residence,
whither the weary toiler may betake himself
periodically (or, more correctly, in relays)
for purposes of refreshment or repose.
Ian Hay, *The First Hundred Thousand*
(1916)

4 The boast of every good battalion was that
it had never lost a trench.
Robert Graves, *Goodbye to All That* (1929)

5 (Of the First World War) When all is said
and done, the war was mainly a matter of
holes and ditches.
Siegfried Sassoon, *Memoirs of an Infantry
Officer* (1930)

6 In the trenches your sins found you out.
A.A. Hanbury-Sparrow, *The Land-Locked
Lake* (1932)

7 The British soldier must be driven to digging
himself in the moment he occupies an area,
and not to waste time in sightseeing,
souvenir-hunting and brewing tea.
General Francis Festing, Order to 36th
Division in Arakan, Burma, 1944

8 A slit trench, after all, is the nearest thing
to a grave we'll be in while we're alive. It *is* a
grave!
R.M. Wingfield, *The Only Way Out* (1955)

9 Dig for discipline. Dig to save your skins.
Dig through sand. Dig, if necessary, through
rock. Dig for bloody victory.
Neil McCallum, *Journey with a Pistol*
(1959)

10 The whole conduct of our trench warfare
seemed to be based on the concept that we,
the British, were not stopping in the
trenches for long, but were tarrying awhile
on the way to Berlin and that very soon we
would be chasing Jerry across country. The
result, in the long term, meant that we lived
a mean and impoverished sort of existence
in lousy scratch holes.
George Coppard, *With a Machine Gun to
Cambrai* (1969)

---

331 TRUCE

---

1 When, without a previous understanding,
the enemy asks for a truce, he is plotting.
Sun Tzu, *The Art of War,* IX (*c.*490 BC)

2 See here my friends and loving countrymen;
This token serveth for a flag of truce
Betwixt ourselves and all our followers:
So help me God, as I dissemble not!
William Shakespeare, *Henry VI,* Part 1,
III, i

3 Our bugles sang truce, for the night-cloud
                 had lower'd,
And the sentinel stars set their watch in the
                 sky;
And thousands had sunk on the ground
                 overpower'd,
The weary to sleep, and the wounded to die.
Thomas Campbell, 'The Soldier's Dream',
Poems (1803)

4 If you are so uncharitable as to refuse me a
truce as requested, then you may do as you
please. I shall not surrender alive. Therefore
bombard as you please.
General Piet Cronje, message to Lord
Kitchener before the Battle of Paardeberg,
1900

5 Just imagine it, English, Scots, Irish,
Prussians, Württemburgers in a chorus. I
wrote a report on the whole fantastic
episode and ended by saying that if I had
seen it on film I would have sworn it was a
fake.
Captain Sir John Hulse, letter to his mother
after the Christmas Day truce on the
Western Front, 1914, quoted in Lamb,
Mutinies (1978)

6 (Of the Christmas Day truce on the Western
Front, 1914) A little human puntuation
mark in our lives of cold and humid hate.
Bruce Bairnsfather, *Bullets and Billets*
(1917)

7 (Of the truce on the Western Front, before
the Armistice, 1918) It gave us an inkling of
the happiness and purity implied in the
word *peace.*
Ernst Jünger, *The Storm of Steel* (1929)

8 That curious week's holiday which the war
had taken which had been so false that they
remembered it only as a phenomenon.
William Faulkner, *A Fable* (1954)

9 As they watched the German trenches,
Something moved in No Man's Land,
And through the dark there came a soldier
Carrying a white flag in his hand.
Mike Harding, 'Christmas 1914', *Oxford
Book of Traditional Verse* (1983)

332 UNIFORMS

1 When a common soldier is civil in his quarters, his red coat procures him a degree of respect.
Samuel Johnson, quoted in Boswell, *The Life of Samuel Johnson* (1791)

2 A man becomes the creation of his uniform.
Napoleon I, *Maxims* (1804–15)

3 I think it is indifferent how a soldier is clothed, provided it is in a uniform manner; and that he is forced to keep himself clean and smart, as a soldier ought to be.
Duke of Wellington, letter to Lt-Col Henry Torrens, 1811

4 A good uniform must work its way with the women, sooner or later.
Charles Dickens, *The Pickwick Papers* (1837)

5 The better you dress a soldier, the more highly he will be thought of by women.
Field Marshal Viscount Wolseley, *The Soldier's Pocket Book* (1869)

6 Yes, makin' mock o' uniforms that guard
 you while you sleep
Is cheaper than them uniforms, an' they're
 starvation cheap;
An' hustlin' drunken soldiers when they're
 goin' large a bit
Is five times better business than parading in
 full kit.
Rudyard Kipling, 'Tommy', *Barrack-Room Ballads* (1892)

7 Someone had said he'd look a god in kilts.
Wilfred Owen, 'Disabled'

8 Every night we carefully soap the insides of our trouser-creases, wet the outsides, and we obtain smartness by laying the damp garments on our mattresses and sleeping on them.
Stephen Graham, *A Private in the Guards* (1919)

9 A sojer's life is on'y gloryous in times of peace. Thin he can wear his good clothes with th' goold lace on thim, an' sthrut in scarlet an' blue through the sthreets.
Finley Peter Dunne, *Mr Dooley on Making a Will* (1919)

10 The secret of a uniform was to make a crowd solid, dignified and impersonal: to give it the singleness and tautness of an upstanding man.
T.E. Lawrence, *The Revolt in the Desert* (1927)

11 It is proverbial that well dressed soldiers are usually well behaved soldiers.
John A. Lejeune, *Reminiscences of a Marine* (1930)

12 Soldiers, in the main, are not impressive clad in the drabness, the futilities of civilian dress.
Joseph Hergesheimer, *Sheridan* (1931)

13 A soldier's body becomes a stock of accessories that are no longer his property.
Antoine de Saint-Exupéry, *Flight to Arras* (1942)

14 Armies have long sought to confer specific abilities on their soldiers by dressing them in a special way.
Richard Holmes, *Firing Line* (1985)

333 WAR

1 War is a fearful thing, but not so fearful that we should submit to anything in order to avoid it.
Polybius, *Histories*, IV

2 Lastlie stode warre in glittering armes yclad,
With visage grim sterne lokes and blacklie
 hued;
In his right hand a naked sword he had
That to the hiltes was all with blood imbrud
And in his left that kinges and kingdomes
 rewed
Famine and fire he held and therwithall
He rased townes and threw doune towres
 and all.
Thomas Sackville, 'A Mirror for Magistrates' (1563)

3 To have prepared for a war before we had entered into it had been good, for as there is nothing we can undertake but we must provide for it, so there is more need for a preparation for war, than anything in the world, for there is no action so great as war.
Sir Edward Cecil, letter to King Charles I, 1626

4 War is the trade of kings
John Dryden, *King Arthur*, II, ii (1691)

5   For I must go where lazy Peace,
    Will hide her drouzy head;
    And, for the sport of Kings, encrease
    The number of the Dead.
    Sir William Davenant, 'The Souldier going
    to the Field' (1693)

6   Soldiers, prepare! Our cause is Heaven's
                            cause;
    Soldiers, prepare! Be worthy of our cause:
    Prepare to meet our fathers in the sky:
    Prepare, O troops, that are to fall to-day!
    Prepare, prepare.
    William Blake, 'A War Song to
    Englishmen', *Poetical Sketches* (1783)

7   Some seek diversion in the tented field,
    And make the sorrows of mankind their
                            sport.
    But war's a game, which, were their subjects
                            wise,
    Kings should not play at.
    William Cowper, *The Task* (1785)

8   War is the statesman's game, the priest's
                            delight,
    The lawyer's jest, the hired assassin's trade,
    And, to those royal murderers, whose mean
                            thrones
    Are bought by crimes of treachery and gore,
    The bread they eat, the staff on which they
                            lean.
    Percy Bysshe Shelley, *Queen Mab* (1813)

9   The man who delights in war is a *madman*;
    I would put him in the thick of it for just one
    day, and he would then know a little of
    what war to the knife means.
    Thomas Gowing, *A Soldier's Experience*
    (1896)

10 We are too much inclined to think of war as
    a matter of combats, demanding above all
    things physical courage. It is really a matter
    of fasting and thirsting; of toiling and
    waking; of lacking and enduring; which
    demands above all things moral courage.
    Sir John Fortescue, *A History of the British
    Army* (1899–1912)

11 Now most soldiers have no experience of
    war; and to assume that those who have are
    therefore qualified to legislate for it, is as
    absurd as to assume that a man who has
    been run over by an omnibus is thereby
    qualified to draw up wise regulations for the
    traffic of London.
    George Bernard Shaw, *John Bull's Other
    Island* (1904)

12 People always make war when they say
    they love peace.
    D.H. Lawrence, 'Peace and War', *Last
    Poems* (1933)

13 In war, whichever side may call itself the
    victor, there are no winners, but all are
    losers.
    Neville Chamberlain, speech at Kettering,
    1938

14 Death's a bastard
    Keeps hitting back.
    But war's a war
    Lady in black.
    Alun Lewis, 'Lady in Black', *Raider's Dawn*
    (1942)

15 One of the essential experiences of war is
    never being able to escape from disgusting
    smells of human origin.
    George Orwell, 'Looking Back on the
    Spanish Civil War' (1943)

16 (Of the English) War was a first-class sport;
    we liked it, we were good at it.
    Marjorie Ward, *The Blessed Trade* (1971)

17 The professional soldier tends to regard war
    as the culmination of his peacetime training,
    and the justification for it.
    Jock Haswell, *Citizen Armies* (1973)

334 WAR CORRESPONDENTS

1   Those newly invented curse to armies who
    eat the rations of the fighting man and do
    not work at all.
    Field Marshal Viscount Wolseley, *The
    Soldier's Pocket Book* (1869)

2   (Of the Crimean War) I could not tell lies to
    'make things pleasant'.
    William Howard Russell, *The British
    Expedition to the Crimea* (1877)

3   Get out of my way, you drunken swabs!
    Field Marshal Earl Kitchener, to the
    accompanying press corps during the Sudan
    campaign, 1898

4   (Of the Sudan campaign) The British public
    likes to read sensational news, and the best
    war correspondent is he who can tell the
    most thrilling lies.
    Field Marshal Earl Haig, letter to his sister
    Henrietta, 1898

5 The first casualty when war comes is truth.
Senator Hiram Johnson, speech to
Congress, 1917

6 (Of the Russian Revolution) In the struggle
my sympathies were not neutral. But in
telling the story of those great days I have
tried to see events with the eye of a
conscientious reporter, interested in setting
down the truth.
John Reed, *Ten Days that Shook the World*
(1919)

7 Even the enormous, impregnable stupidity
of our High Command in all matters of
psychology was penetrated by a vague
notion that a few 'writing fellows' might be
sent out with permission to follow the
armies in the field, under the strictest
censorship, in order to silence the clamour
for more news.
Philip Gibbs, *Realities of War* (1920)

8 (Of the Spanish Civil War) Early in life I
have noticed that no event is ever correctly
reported in a newspaper, but in Spain, for
the first time, I saw newspaper reports
which did not bear any relation to the facts,
not even the relationship which is implied in
an ordinary lie.
George Orwell, 'Looking Back on the
Spanish Civil War' (1943)

9 I wouldn't tell the people anything until the
war is over, and then I'd tell them who won.
Anon., attributed to an American military
censor of the Second World War

10 (Of the Second World War) We were
cheerleaders. I suppose there wasn't an
alternative at the time. It was total war. But,
for God's sake, let's not glorify our role. It
wasn't good journalism. It wasn't
journalism at all.
Charles Lynch, quoted in Knightley, *The
First Casualty* (1975)

11 I have never believed that correspondents
who move in and out of the battle area,
engage in privileged conversations with
commanders and troops, and who have
access to a public platform, should engage
in criticism of command decisions or of
commanders while the battle is in progress.
Ed Murrow, *In Search of Light* (1967)

12 It may well be that between press and
officials there is an inherent, built-in conflict
of interest. There is something to be said for
both sides, but when the nation is at war
and men's lives are at stake, there should be
no ambiguity.
General William C. Westmoreland, *A
Soldier Reports* (1976)

13 (Of the Vietnam War) Conventional
journalism could no more reveal this war
than conventional firepower could win it, all
it could do was take the most profound
event of the American decade and turn it
into a communications pudding, taking its
most obvious undeniable history and
making it into a secret history.
Michael Herr, *Dispatches* (1977)

14 I counted them all out and I counted them
all back.
Brian Hanrahan, BBC Television news
bulletin on the Falklands War, 1982

15 (Of the civil war in Lebanon) If reporters
want to take sides in Lebanon then they
should give up journalism and join the army
of their choice. The fact of the matter is that
there are no good guys in Lebanon —
they're all bad. if you don't think that then
you should fly home on the next plane.
Robert Fisk, quoted in *The Times: Past,
Present and Future* (1985)

335 WAR LITERATURE

1 Who shall record the glorious deeds of the
soldier whose lot is numbered with the
thousands in the ranks who live and die and
fight in obscurity?
Private Wheeler, letter of 1813, quoted in
Liddell Hart (ed.), *The Letters of Private
Wheeler* (1951)

2 Oh! shrink not thou, reader! Thy part's in it,
    too;
Has not thy praise made the thing they go
        through
Shocking to read of, but noble to do?
Leigh Hunt, 'Captain Sword and Captain
Pen' (1835)

3 To men of a sedate and mature spirit, in
whom is any knowledge or mental activity,
the detail of battle becomes insupportably
tedious and revolting.
Ralph Waldo Emerson, *War* (1849)

4 Amid the crash of shells and the whistle of

bullets, the cheers and the dying cries of comrades, the sense of personal danger, the pain of wounds, and the consuming passion to reach an enemy, he must be an exceptional man who is cool enough and curious enough to be looking serenely about him for what painters call 'local colour'.
J.W. Wightman, on the Battle of Balaclava, quoted in *Nineteenth Century*, 1890

5 My argument is that War makes rattling good history; but Peace is poor reading.
Thomas Hardy, *The Dynasts* (1904–08)

6 Damn your writing — mind your fighting.
General Viscount Lake, quoted in Pearse, *Life and Military Services of Viscount Lake* (1908)

7 Above all, this book is not concerned with Poetry.
The subject of it is War, and the Pity of War.
The Poetry is in the Pity.
Wilfred Owen, Preface to *Poems* (1920)

8 We in our haste can only see the small components of the scene
We cannot tell what incidents will focus on the final screen.
Donald Bain, 'War Poets', Hamilton, *The Poetry of War 1939–1945* (1972)

9 It is the logic of our times,
No subject for immortal verse —
That we who lived by honest dreams
Defend the bad against the worse.
C. Day Lewis, 'Where are the War Poets?' (1943)

10 No, I cannot write the poem of war,
Neither the colossal dying nor the local scene,
A platoon asleep and dreaming of girl's warmth
Or by the petrol-cooker scraping out a laughter.
Robert Conquest, 'Poem in 1944'

11 I am the man who groped for words and found
An arrow in my hand.
Sidney Keyes, 'War Poet', *Collected Poems* (1945)

12 And there's the outhouse poet, anonymous:
Soldiers who wish to be a hero
Are practically zero
But those who wish to be civilians
Jesus they run into millions.
Norman Rosten, *The Big Road* (1946)

13 I wanted people to find in my poems the truth of what it had been like to be an American infantry soldier.
Louis Simpson, quoted in Hamilton, *The Poetry of War, 1939–1945* (1972)

14 War stories aren't really anything more than stories about people anyway.
Michael Herr, *Dispatches* (1977)

---

336 WEAPONS
See also 273. BAYONETS, 302. MACHINE GUNS, 320. RIFLES, 326. SWORDS

---

1 Cannon and fire-arms are cruel and damnable machines; I believe them to have been the direct suggestion of the devil.
Martin Luther, *Table Talk* (1569)

2 The instruments of battle are valuable only if one knows how to use them.
Charles Ardant du Picq, *Battle Studies* (1870)

3 The man is the first weapon of battle.
Charles Ardant du Picq, *Battle Studies* (1870)

4 When confidence is placed in superiority of material means, valuable as they are against an enemy at a distance, it may be betrayed by the actions of the enemy. If he closes with you in spite of your superiority in means of destruction, the morale of the enemy mounts with your loss of confidence.
Charles Ardant du Picq, *Battle Studies* (1870)

5 He has made his weapons his gods. When his weapons win he is defeated himself.
Rabindranath Tagore, *Stray Birds* (1916)

6 There can never be too many guns, there are never enough of them.
Marshal Ferdinand Foch, *Precepts* (1919)

7 We can do without butter, but, despite all our love of peace, not without arms. One cannot shoot with butter, but with guns.
Josef Goebbels, speech in Berlin, 1936

8 It is war that shapes peace and armament that shapes war.
J.F.C. Fuller, *Armament and History* (1945)

9 There are two universal and important weapons of the soldier which are often

overlooked — the boot and the spade. Speed and length of marching has won many victories; the spade has saved many defeats and gained time for victory.
Field Marshal Earl Wavell, *The Good Soldier* (1945)

10 It never has made and never will make any sense trying to abolish any particular weapon of war. What we have to abolish is war.
Sir John Slessor, *Strategy for the West* (1954)

11 Everything that is shot or thrown at you, or dropped on you in war is unpleasant, but of all horrible devices, the most terrifying is the land mine.
Field Marshal Viscount Slim, *Unofficial History* (1959)

12 Today the expenditure of billions of dollars every year on weapons, acquired for the purpose of making sure we never need to use them, is essential to keeping the peace.
John F. Kennedy, speech at the American University, Washington, 1963

13 Grenades were one of the most fearfully respected and accident-prone tricky instruments an infantryman had to deal with. They were often as likely to hurt your own people or yourself as the enemy.
James Jones, *WWII* (1977)

14 Losing guns in battle has always been more bitterly regretted than the weapons' real military value might suggest, and artillerymen who might have run away to fight another day have often stood fast about their silent guns, selling their lives dearly with handspike and hammer.
Richard Holmes, *Firing Line* (1985)

337 WEST POINT

1 It but rarely happens that a graduate from West Point is not a gentleman in his deportment, as well as soldier in his education.
Colonel Archibald Henderson, letter to the Secretary of the United States Navy, 1823

2 The standards for the American Army will be those of West Point. The rigid attention, the upright bearing, attention to detail,

uncomplaining obedience to instruction, required of the cadet, will be required of every officer and soldier of our armies in France.
General John F. Pershing, General Order to the US forces in France, 1917

3 In my dreams I hear again the crash of guns the rattle of musketry, the strange, mournful mutter of the battlefield. But in the evening of my memory, always I come back to West Point.
General Douglas MacArthur, address at the US Military Academy, West Point, 1962

4 When the going gets tough, the tough get going.
Anon., cadet saying, West Point

5 Duty, honor, country.
Motto of the United States Military Academy, West Point

338 WOUNDS

1 A wound is nothing, be it ne'er so deep;
Blood is the gold of war's rich livery.
Christopher Marlowe, *Tamburlaine the Great* Part 2, II, ii

2 The history of a soldier's wound beguiles the pain of it.
Laurence Sterne, *Tristram Shandy* (1759–67)

3 The broken soldier, kindly bade me stay.
Sat by his fire, and talk'd the night away;
Wept o'er his wounds, or tales of sorrow
        done,
Shoulder'd his crutch, and show'd how
        fields were won.
Oliver Goldsmith, 'The Deserted Village' (1770)

4 Ben Battle was a soldier bold,
And used to war's alarms:
But a cannon-ball took of his legs,
So he laid down his arms!
Thomas Hood, 'Faithless Nelly Gray'

5 (Of the Battle of Albuera) Every individual most nobly did his duty; and it was observed that our dead, particularly the 57th Regiment, were lying as they fought, in ranks, and every wound was in front.
Lt-Gen Viscount Beresford, despatch to Wellington, 1811

6 A soldier wounded in action always thinks the affair lost and imagines it to have been a very bloody fight.
Count Leo Tolstoy, *Tales of Army Life* (1855)

7 (Of the Crimean War) Some of the men's wounds were left so long unattended to that those men who were able would do a good turn by picking the maggots from them. This fact is almost too bad to print, but I mention it as one of the horrors of war.
Sergeant Taffs, quoted in Small, *Told From the Ranks* (1897)

8 When a soldier's resistance to fear has been lowered by sickness or by a wound the balance has been tilted against him and his control is in jeopardy at any rate for a time. The wounded soldier has just visualised danger in a new and very personal way.
Lord Moran, *The Anatomy of Courage* (1945)

9 The wounded, who could not be brought in, had crawled into shell holes, wrapped their waterproof sheets round them, taken out their Bibles, and died like that.
Gerald Brenan, *A Life of One's Own* (1962)

10 The late-twentieth-century soldier does not expect to be left to die of his wounds on the battlefield.
John Keegan, *The Face of Battle* (1976)

11 I realised vividly now that the real horrors of war were to be seen in hospitals, not on the battlefield.
Lt-Gen Sir John Glubb, *Into Battle: A Soldier's Diary of the Great War* (1978)

12 Like most people I had not fully realised that the horror of war is wounds, not death.
Captain C.S. Stormont Gibbs, *From the Somme to the Armistice* (1986)

V
LAST POST

339 CASUALTIES

1 How are the mighty fallen in the midst of
battle!
II Samuel 1:25

2 If bloody slaughter is a horrible sight, then
that is a ground for paying more respect to
war.
Karl von Clausewitz, *On War* (1832)

3 They are bringing him down,
He looks at me wanly.
The bandages are brown,
Brown with mud, red only —
But how deep a red!
Robert Nichols, 'Casualty', *Ardours and
Endurances* (1917)

4 One dies of war like any old disease.
Wilfred Owen, 'A Terre, Being the
Philosophy of Many Soldiers' (1917)

5 One cannot see these ragged and putrid
bundles of what once were men without
thinking of what they were — their
cheerfulness, their courage, their idealism,
their love for their dear ones at home.
Lt-Gen Sir John Glubb, 1917, *Into Battle; A
Soldier's Diary of the Great War* (1978)

6 The most fatal heresy in war, and, with us,
the most rank, is the heresy that battles can
be won without heavy loss.
General Sir Ian Hamilton, *Gallipoli Diary*
(1920)

7 None saw their spirits' shadow shake the
grass.
Or stood aside for the half used life to pass

Out of these doomed nostrils and
the doomed mouth,
When the swift iron burning bee
Drained the wild honey of their youth.
Isaac Rosenberg, 'Dead Man's Dump',
Poems (1922)

8 Among the wire lay rows of khaki figures, as
they had fallen to machine-guns on the
crest, thick as the sleepers in the Green Park
on a summer Sunday evening.
Charles Carrington, *A Subaltern's War*
(1929)

9 Let me not mourn for the men who have
died fighting, but rather, let me be glad that
such heroes have lived.
General George S. Patton, 'Soldier's
Prayer', quoted in Semmes, *Portrait of
Patton* (1955)

10 Live and let live.
No matter how it ended,
These lose and, under the sky,
Lie befriended.
John Pudney, 'Graves: El Alamein',
Collected Poems (1957)

11 It is always rather a pitiful business seeing
men you have shot, even enemies in war.
Field Marshal Viscount Slim, *Unofficial
History* (1959)

12 To a foot-soldier, war is almost entirely
physical. That is why some men, when they
think about war, fall silent. Language seems
to falsify physical life and to betray those
who have experienced absolutely — the
dead.
Louis Simpson, quoted in Hamilton, *The
Poetry of War 1939–1945* (1965)

13 A casualty is a man blown to pieces, disintegrated, nothing left of him but a name on a war memorial.
John Terraine, *The Smoke and the Fire* (1980)

14 Sensitivity over casualties can be self-defeating unless balanced by courageous judgement.
General Sir David Fraser, *And We Shall Shock Them* (1983)

340 DEATH

1 The household knew those whom it sent to war; but instead of the men an urn and a handful of ashes alone returned.
Aeschylus, *The Agamemnon,* trs. Smyth

2 Few, few shall part where many meet,
The snow shall be their winding-sheet,
And every turf beneath our feet,
Shall be a soldier's sepulchre.
Thomas Campbell, 'Hohenlinden', *Poems* (1803)

3 For the Angel of Death spread his wings on the blast,
And breathed in the face of the foe as he pass'd.
Lord Byron, *Don Juan,* I (1819)

4 (Of the Battle of Waterloo) Oh, do not *congratulate* me. I have lost all my dearest friends.
Duke of Wellington, remark to the Hon. Mrs Wellsley Pole, quoted in Bagot, *Links with the Past* (1901)

5 Bullets are like letters; when someone writes my address on its envelope, I'll be getting it.
General Emilio Vidal Mola, 1912

6 We are the Dead. Short days ago
We lived, felt dawn, saw sunset glow,
Loved and were loved, and now we lie
In Flanders fields.
John McCrae, 'In Flanders Fields' (1914)

7 When you see millions of the mouthless dead
Across your dreams in pale battalions go,
Say not soft things as other men have said,
That you'll remember. For you need not so.
Charles Hamilton Sorley, 'When you see millions of the mouthless dead' (1914)

8 If I should die, think only this of me:
That there's some corner of a foreign field
That is for ever England.
Rupert Brooke, 'The Soldier', *1914 and Other Poems* (1915)

9 Jolly young Fusiliers, too good to die.
Robert Graves, 'The Last Post' (1916)

10 What passing bells for those who die as cattle?
Only the monstrous anger of the guns.
Wilfred Owen, 'Anthem for Doomed Youth' (1917)

11 Here dead we lie because we did not choose
To live and shame the land from which we sprung.
Life, to be sure, is nothing much to lose;
But young men think it is, and we were young.
A.E. Housman, 'Here Dead We Lie', *Last Poems* (1922)

12 I was looking straight at him as the bullet struck him and was profoundly affected by the remembrance of his face, though at the time I hardly thought of it. He was alive, and then he was dead, and there was nothing human left in him. He fell with a neat round hole in his forehead and the back of his head blown out.
Charles Carrington, *A Subaltern's War* (1929)

13 Shell-twisted and dismembered, the Germans maintained the violent attitudes in which they had died.
Siegfried Sassoon, *Memoirs of an Infantry Officer* (1930)

14 Word came to him in the bullet shower
that he should be a hero briskly,
and he was that while he lasted
but it wasn't much time he got.
Sorley MacLean, 'Curaidhean' ('Heroes'), *Selected Poems* (1977)

15 He had been killed while climbing up the steep bank of the Bluff, and had one foot raised and a hand stretched out to pull himself up. By some miracle he remained in the same identical position. Except for the green colour of his face and hand, one

would never have believed that he was dead.
Lt-Gen Sir John Glubb, *Into Battle: A Soldier's Diary of the Great War* (1978)

---

341 DEFEAT

1 *Vae victis.* Woe to the vanquished.
Livy, *Histories,* V

2 A defeated ruler should never be spared.
Stendhal, *Life of Napoleon* (1818)

3 We are not interested in the possibilities of defeat.
Queen Victoria, discussing the Boer War with A.J. Balfour, 1899

4 A beaten general is disgraced for ever.
Marshal Ferdinand Foch, *Precepts* (1919)

5 The winner is asked no questions — the loser has to answer for everything.
General Sir Ian Hamilton, *Gallipoli Diary* (1920)

6 Defeat is a thing of weariness, of incoherence, of boredom. And above all, futility.
Antoine de Saint-Exupéry, *Flight to Arras* (1942), trs. Galantière

7 Man in war is not beaten, and cannot be beaten, until he owns himself beaten.
Captain Sir Basil Liddell Hart, *Thoughts on War* (1944)

8 Fearful are the convulsions of defeat.
Winston S. Churchill, *The Gathering Storm* (1948)

9 The quickest way of ending a war is to lose it.
George Orwell, 'Second Thoughts on James Beecham', *Shooting an Elephant* (1950)

10 A defeated man does not make a good philosopher.
Jean Dutourd, *The Taxis of the Marne* (1957)

11 A battle is not lost until the commander believes it to be lost.
Jock Haswell, *Citizen Armies* (1973)

12 In the cold accountancy of war and history there may be headlines to be extracted from defeat, but there is not virtue.
Max Hastings, *The Korean War* (1987)

---

342 EXECUTIONS

1 The most honourable death for a delinquent soldier is beheading.
Sir James Turner, *Pallas Armata* (1683)

2 Nothing can be more solemn or impressive than a military execution.
J. MacMullen, *Camp and Barrack Room* (1846)

3 They've taken of 'is buttons off and cut 'is stripes away,
And they're hangin' Danny Deever in the mornin'.
Rudyard Kipling, 'Danny Deever', *Barrack-Room Ballads* (1892)

4 Blindfolded, when the dawn was grey,
He stood there in a place apart,
The shots rang out and down he fell,
An English bullet in his heart.
Winifred M. Letts, 'The Deserter', 1915, quoted in Giddings, *The War Poets* (1988)

5 I could not look on Death, which being known,
Men led me to him, blindfold and alone.
Rudyard Kipling, 'The Coward', *Epitaphs of the War* (1919)

6 The death penalty for cowardice was a deterrent which prevented many a man from being a coward, and possibly saved many lives through the deterrent effect being powerful enough to prevent the occurrence of cowardice.
Thomas Shaw, speech in the House of Commons, 1928

7 It is not fair to take a man from a farm or factory, clap a tin hat on his head, and then shoot him if his nerve fails.
Ernest Thurtle, speech on the House of Commons, 1928

8 The only deterrent for the man who will wilfully behave in such a way as to endanger the lives of his own comrades in order to avoid the risk to his own life, is the knowledge that while his comrades might possibly incur death at the hands of the enemy which will be a glorious and honourable death, he, if convicted of one of these offences, will die a death which is dishonourable and shameful.
Field Marshal Viscount Allenby, speech in the House of Lords, 1928

9 How easy for the generals living in luxury
well back in their chateaux to enforce the
death penalty and with the stroke of a pen
sign some poor wretch's death warrant.
Maybe of some poor, half-witted farm
yokel, who once came forward of his own
free will without being fetched.
M. Evans, *Going Across* (1952)

10 At dawn a bird will waken me
Unto my place among the kings.
Francis Ledwidge, 'After Court Martial',
Complete Poems (1974)

11 It was contended by the fighting soldiers
that the absence of the death penalty for
desertion placed an intolerable burden on
men's courage and determination by
removing any comparable fear of the
consequences of failure.
General Sir David Fraser, *And We Shall
Shock Them* (1983)

---

343 KILLING

---

1 Saul hath slain his thousands and David his
ten thousands.
1 Samuel, 18: 7

2 Next Erymas was doomed his fate to feel;
His opened mouth received the Cretan steel;
Beneath the brain the point a passage tore,
Crashed the thin bones, and drowned the
                              teeth in gore.
His mouth, his eyes, his nostrils, pour a
                              flood;
He sobs his soul out in the gush of blood.
Alexander Pope, *Iliad,* XVI (1715)

3 War its thousands slays, Peace its ten
thousands.
Beilby Porteous, *Death* (1759)

4 To kill a human being is, after all, the least
injury you can do him.
Henry James, *My Friend Bingham* (1867)

5 It is nice to put a sword or a lance through a
man; they are just like old hens, they just say
'quar'.
Private Rawding, 21st Lancers, letter to his
parents describing the Battle of Omdurman,
1898, quoted in Emery, *Marching Over
Africa* (1986)

6 Now and then I caught in a man's eye the
curious gleam which comes from the joy of
shedding blood — the mysterious impulse
which, despite all the veneer of civilisation,
still holds its own in a man's nature,
whether he is killing rats with a terrier,
rejoicing in the prize fight, playing a salmon
or potting Dervishes.
Ernest Bennett, in the *Westminster Gazette,*
1898

7 Nothing is ever done in this world until men
are prepared to kill one another if it is not
done.
George Bernard Shaw, *Major Barbara,* III
(1907)

8 I saw him stab
and stab again
a well-killed Boche.
Herbert Read, 'The Happy Warrior', *Naked
Warriors* (1919)

9 A man cannot change his feelings again
during the last rush with a veil of blood
before his eyes. He does not want to take
prisoners but to kill.
Ernst Junger, *The Storm of Steel* (1929)

10 When you have to kill a man it costs nothing
to be polite.
Winston S. Churchill, on the declaration of
war against Japan, 1941, quoted in *The
Grand Alliance,* III (1950)

11 Kill a man and you are a murderer. Kill
millions of men and you are a conqueror.
Kill everyone and you are a god.
Jean Rostand, *Pensées d'un Biologiste*
(1955)

12 Killing
Is the ultimate simplification of life.
Hugh MacDiarmid, 'England's Double
Knavery', *The Clyack-Sheaf* (1969)

13 The basic aim of a nation at war in
establishing an image of the enemy is to
distinguish as sharply as possibly the act of
killing from the act of murder by making the
former into one deserving of all honour and
praise.
J. Glenn Gray, *The Warriors: Reflections on
Men in Battle* (1970)

14 The morality of killing is not something
with which the professional soldier is
usually thought to trouble himself.
John Keegan, *The Face of Battle* (1976)

344 LAST POST

1 No longer hosts encountering hosts
Shall crowds of slain deplore;
They hang the trumpet in the hall,
And study war no more,
Scottish Psalter, Paraphrase XVIII

2 Nothing is here for tears, nothing to wail
Or knock the breast; no weakness or
                              contempt.
Dispraise or blame; nothing but well and
                              fair,
And what may quiet us in a death so noble.
John Milton, 'Samson Agonistes' (1671)

3 And all the trumpets sounded for him on the
other side.
John Bunyan, *The Pilgrim's Progress* (1678)

4 How sleep the brave, who sink to rest
By all their Country's wishes blest.
William Collins, 'Written after the Battle of
Fontenoy' (1745)

5 Dool and wae for the order sent our lads to
                              the Border!
The English, for once, by guile wan the day;
The Flowers of the Forest, that foucht aye
                              the foremost —
The prime o' our land — are cauld in the
                              clae.
Jean Elliot, 'The Flowers of the Forest',
Herd's Scottish Songs (1776)

6 Soldier, rest! thy warfare is o'er,
Dream of fighting fields no more;
Sleep the sleep that knows not breaking,
Morn of toil, nor night of waking.
Sir Walter Scott, *The Lady of the Lake*, I
(1810)

7 Wrap me up in my tarpaulin jacket,
And say a poor buffer lies low,
And six stalwart lancers shall carry me
With steps solemn, mournful and slow.
George J. Whyte-Melville, 'The Tarpaulin
Jacket', *Songs and Verses* (1869)

8 East and west on fields forgotten
Bleach the bones of comrades slain,
Lovely lads and dead and rotten;
None that go return again.
A.E. Housman, *A Shropshire Lad* (1896)

9 They shall grow not old, as we that are left
grow old:
Age shall not weary them, nor the years
condemn.

At the going down of the sun and in the
                              morning
We will remember them.
Laurence Binyon, 'For the Fallen', *The
Times* (1914)

10 Blow out, you bugles, over the rich Dead!
Rupert Brooke, 'The Dead', *1914 and
Other Poems* (1915)

11 When the Last Post is blown
And the last volley fired,
When the last sod is thrown,
And the last foe retired,
And the last bivouac is made under the
                              ground —
Soldier, sleep sound.
Joseph Lee, 'Ballads of Battle' (1916)

12 Not in the Abbey proudly laid
Find they a place or part;
The gallant lads of the old brigade,
They sleep in old England's heart.
Frederic Edward Weatherley, 'The Old
Brigade' (1919)

13 O Valiant Hearts, who to your glory came
Through dust of conflict and through battle
                              flame;
Tranquil you lie, your knightly virtue
                              proved,
Your memory hallowed in the land you
                              loved.
Sir John Stanhope Arkwright, 'O Valiant
Heart', *The Supreme Sacrifice* (1919)

14 They shall receive a Crown of Glory that
fadeth not away.
Rudyard Kipling, inscription for the Menin
Gate Memorial at Ypres, Belgium, 1927

15 Here dead we lie because we did not choose
To live and shame the land from which we
                              sprung.
Life to be sure, is nothing much to lose;
But young men think it is, and we were
                              young.
A.E. Housman, 'Here Dead We Lie', *More
Poems* (1936)

16 No man outlives the grief of war
Though he outlive its wreck;
Upon the memory a scar
Through all his years will ache.
William Soutar, 'The Permanence of Young
Men' (1940)

17 When you go home,
Tell them of us, and say —

For your tomorrow,
We gave our today.
J. Maxwell Edmonds, epitaph on 2nd
Division Memorial, Kohima, Burma, 1944

---

345 PATRIOTISM

---

1 *Dulce et decorum est pro patria mori.* It is a
   sweet and seemly thing to die for one's
   country.
   Horace, *Odes,* III, ii

2 It is better for us to die in battle than to
   behold the calamities of our people, and our
   sanctuary.
   I Maccabees 3:59

3 For as long as there shall be but one hundred
   of us remain alive, we will never consent to
   subject ourselves to the dominion of the
   English.
   Bernard de Linton, letter to Pope John
   XXII, 1320, also known as the Declaration
   of Arbroath

4 The best fortress is to be found in the love of
   the people.
   Niccolo Machiavelli, *The Prince* (1513)

5 Dying for one's country is not a bad fate —
   you become immortal through a fine death.
   Pierre Corneille, *Le Cid* (1637)

6 I regret that I have but one life to lose for my
   country.
   Captain Nathan Hale, last words before his
   execution, 1776

7 Patriotism is the last refuge of a scoundrel.
   Samuel Johnson, quoted in Boswell, *Life of
   Johnson* (1791)

8 Breathes there the man with soul so dead,
   Who never to himself hath said,
   'This is my own, my native land!'
   Sir Walter Scott, *The Lay of the Last
   Minstrel,* VI (1805)

9 'Qui procul hinc', the legend's writ —
   The frontier is far away —
   'Qui ante diem periit:
   Sed miles, sed pro patria,'
   (Who died far from here, before his time,
             but as a soldier, and for his
             country.)
   Sir Henry Newbolt, 'Clifton Chapel' (1897)

10 I realise that patriotism is not enough. I

must have no hatred or bitterness toward
any one.
Nurse Edith Cavell, last words before her
execution by a German firing squad, 1915

11 In the beginning of a change, the patriot is a
   scarce man, and brave and hated and
   scorned. When his cause succeeds, the timid
   join him, for then it costs nothing to be a
   patriot.
   Mark Twain, *Notebooks* (1935)

12 For Soviet patriots homeland and
   communism are fused into one inseparable
   whole.
   Vyacheslav M. Molotov, speech to the
   Supreme Soviet, 1939

13 The Nation today needs men who think in
   terms of service to their country, and not in
   terms of their country's debt to them.
   General Omar N. Bradley, *Military Review*
   (1948)

14 Patriotism is not a short and frenzied
   outburst of emotion, but the tranquil and
   steady dedication of a lifetime.
   Adlai Stevenson, speech in New York, 1952

15 Ask not what your country can do for you
   — ask what you can do for your country.
   John F. Kennedy, Inaugural Address as
   President, 1961

---

346 PYRRHIC VICTORY

---

1 It was a Cadmeian sort of victory with more
   loss than gain.
   Herodotus, *Histories,* I, 166, trs. de
   Selincourt

2 If we are victorious in one more battle with
   the Romans, we shall be utterly ruined.
   Pyrrhus of Epirus, after the Battle of
   Heraclea, 280 BC, quoted in Plutarch, *Lives,*
   trs. Perrin

---

347 RETREAT

---

1 To those who flee comes neither power nor
   glory.
   Homer, *The Iliad,* XV

2 A fine retreat is as good as a gallant attack.
   Baltasar Gracian, *The Art of Worldly Wisdom,* 38 (1647)

3 In all the trade of war, no feat
   Is nobler than a brave retreat.
   Samuel Butler, *Hudibras,* III (1678)

4 A reverse is not a crime when everything has been done to merit a victory.
   Lazare Carnot, letter to General Hoche, after the Battle of Kaiserlauten, 1793

5 Honourable retreats are no ways inferior to brave charges, as having less fortune, more of discipline, and as much valour.
   Maj-Gen Sir William Napier, *Peninsular War,* IV (1810)

6 In a retreat, besides the honour of the army, the loss of life is often greater than in two battles.
   Napoleon I, *Maxims of War* (1831)

7 If we are surrounded we must cut our way out as we cut our way in.
   General Ulysses G. Grant, Order to his troops at the River Belmont, Missouri, 1861

8 The word 'retire' is absolutely forbidden in this division, and some other phrase has to be used where any withdrawal is required.
   Maj-Gen J.G. Legge, Order to 2nd Australian Division, 1916

9 When a heavy field gun trots you may be sure things are pretty bad.
   Field Marshal Sir John French, quoted in Riddell, *War Diary* (1933)

10 I saw young Americans turn and bolt or throw down their arms cursing their government for what they thought was embroilment in a hopeless cause.
   Marguerite Higgins, on the retreat to the Naktong, Korean War, in the *New York Tribune,* 1950

11 Retreat, hell! We're just fighting in another direction.
   Maj-Gen Oliver P. Smith, before the retreat from the Chosin Reservoir, Korean War, 1950

12 The soldier who has been forced to retreat through no fault of his own loses confidence in the higher command; because he has withdrawn already from several positions in succession he tends to look upon retreat as an undesirable but natural outcome of a battle.
   Field Marshal Earl Alexander, *Memoirs* (1962)

---

348 SUICIDE

1 He is truly great who has not only given himself the order to die, but has found the means.
   Seneca, *Ad Lucilium,* Epistle 70

2 While fleeing from an enemy, Fannius killed himself. Is not this, I ask, madness — to die to avoid death?
   Martial, *Epigrams,* II, 80

3 We should not desert from the world's garrison without the express command of he who has placed us here.
   Michel de Montaigne, *Essays,* II, 3 (1580)

4 This life's a fort commited to my trust, Which I must not yield up till it be forced.
   Philip Massinger, *The Maid of Honour,* IV, 3 (1632)

5 Men, like soldiers, may not quit the post.
   Lord Tennyson, *Lucretius,* I (1868)

6 One should die proudly when it is no longer possible to live proudly.
   Friedrich Nietzsche, *Twilight of the Idols* (1889)

7 When you're wounded and left on Afghanistan's plains,
   An' the women come out to cut up what remains,
   Jest roll to your rifle an' blow out your brains
   An' go to your Gawd like a soldier.
   Rudyard Kipling, 'The Young British Soldier', *Barrack-Room Ballads* (1892)

8 With him they buried the muzzle his teeth had kissed,
   And truthfully wrote the mother, 'Tim died smiling.'
   Wilfred Owen, 'S.I.W.'

9 In winter trenches, cowed and glum, With crumps and lice and lack of rum, He put a bullet through his brain, No one spoke of him again.
   Siegfried Sassoon, 'Suicide in the Trenches', *Counter-Attack* (1918)

10 When the nerves break down, there is nothing left but to admit that one can't handle the situation and to shoot oneself.
Adolf Hitler, after the surrender of the German Army at Stalingrad, 1943, quoted in Clark, *Barbarossa: The Russian-German Conflict* (1965)

349 SURRENDER

1 We give up the fort when there's not a man left to defend it.
General Croghan, address to his men at Fort Stevenson, 1812

2 My wounded are behind me and I will never pass them alive.
General Zachary Taylor, on receiving the order to retire during the Battle of Buena Vista, 1847

3 I am tired; my heart is sick and sad. From where the sun now stands, I will fight no more forever.
Chief Joseph, statement to General Nelson Miles, 1877

4 Surrender is essentially an operation by means of which we set about explaining instead of acting.
Charles Péguy, *Les Cahiers de la Quinzaine* (1905)

5 Even barbarians will not surrender their cattle without a struggle.
Louis Botha, on the German surrender, 1918, quoted in Mee, *The End of Order: Versailles 1919* (1980)

6 Helplessness induces hopelessness, and history attests that loss of hope and not loss of lives is what decides the issue of war.
Captain Sir Basil Liddell Hart, *The Real War 1914–1918* (1930)

7 No matter how long this war lasts, Germany will never capitulate. Never will we repeat the mistake of 1918, of laying down our arms at a quarter to twelve.
Adolf Hitler, speech in 1943, quoted in *Table Talk* (1953)

8 To the German Commander: NUTS.
Signed, The American Commander.
Brig-Gen A.C. McAuliffe, message to the German forces on being requested to surrender Bastogne, 1944, quoted on BBC Radio *War Report*, 1944

9 In receiving your surrender I do not recognise you as honourable and gallant men: you will be treated with due but severe courtesy.
General Sir Thomas Blamey, speech accepting the surrender of the Japanese forces in the Dutch East Indies, 1945

10 The surrender of Germany came not with a bang but a whimper.
Alan Moorehead, *Eclipse* (1945)

11 To receive the unconditional surrender of half a million enemy soldiers, sailors and airmen must be an event which happens to few people in the world. I was very conscious that this was the greatest day of my life.
Admiral Earl Mountbatten, Diary, on taking the surrender of the Japanese forces, Singapore, 1945

12 No soldier can claim a right to 'quarter' if he fights to the extremity.
Charles Carrington, *Soldier from the Wars Returning* (1965)

13 A shout of surrender from the darkness of a dugout was too often an invitation to receive a grenade, the wave of an arm from the hatch of a disabled vehicle the signal to unleash a burst of automatic fire.
John Keegan, *The Face of Battle* (1976)

350 VETERANS

1 One who in the past had suffered much in the wars and from the waves; now he slept at peace forgetful of what he had suffered.
Homer, *Odyssey*

2 Before my back was bent I was handsome, My spear was first in battle, it led the attack, Now I am bowed, I am heavy, I am sad.
Anon., 'Lament in Old Age' from *Canu Llywarch Hen* (c.900)

3 Our God and soldiers we alike adore Ev'n at the brink of danger, not before: After deliverance, both alike requited, Our God's forgotten, and our soldiers slighted.
Francis Quarles 'Epigram' (1630)

4   Hacked, hewn with constant service,
         thrown aside,
    To rest in peace and rot in hospitals.
    Thomas Southerne, *The Loyal Brother* (1682)

5   As long as there are a few veterans you can
    do what you like with the rest.
    Marshal Maurice de Saxe, *Mes Rêveries*
    (1757)

6   It is the height of injustice not to pay a
    veteran more than a reservist.
    Napoleon I, *Maxims of War* (1831)

7   All the world over, nursing their scars,
    Sit the old fighting-men broke in the wars —
    Sit the old fighting-men, surly and grim,
    Mocking the lilt of the conquerors' hymn.
    Rudyard Kipling, verse from *Many
    Inventions* (1893)

8   Old soldiers never die,
    Never die, never die,
    Old soldiers never die,
    They only fade away.
    Anon., song of the First World War, 1914–18

9   The nation which forgets its defenders will
    itself be forgotten.
    Calvin Coolidge, speech in New York, 1920

10 Middle-aged men, strenuously as they
    attempt to deny it, are united by a secret
    bond and separated by a mental barrier from
    their fellows who were too old or too young
    to fight in the Great War.
    Charles Carrington, *A Subaltern's War*
    (1929)

11 If we had our time over again we would not
    leave the Army until we were damned well
    kicked out of it.
    Frank Richards, *Old-Soldier Sahib* (1936)

12 It takes a few veterans to leaven a division of
    doughboys.
    General George S. Patton, *War As I Knew It*
    (1947)

351 VICTORY

1   Victory often changes her side.
    Homer, *The Iliad*

2   It is no doubt a good thing to conquer on
    the field of battle, but it needs greater
    wisdom and greater skill to make use of
    victory.
    Polybius, *Histories*

3   They make a desert and they call it peace.
    Tacitus, *Agricola*

4   Victory in war does not depend entirely
    upon numbers or mere courage; only skill
    and discipline will insure it.
    Vegetius, *De Re Militari* (378)

5   He that is taken and put into prison is not
    conquered, though overcome; for he is still
    an enemy.
    Thomas Hobbes, *Leviathan* (1652)

6   The right of conquest has no foundation
    other than the right of the strongest.
    Jean-Jacques Rousseau, *Social Contract*
    (1762)

7   Life's sovereign moment is a battle won.
    Oliver Wendell Holmes, *The Banker's
    Secret* (1850)

8   The god of Victory is said to be one-handed,
    but Peace gives Victory to both.
    Ralph Waldo Emerson, *Journals* (1867)

9   Man does not enter battle to fight, but for
    victory. He does everything he can to avoid
    the first and obtain the second.
    Charles Ardant du Picq, *Battle Studies*
    (1870)

10 The will to conquer is victory's first
    condition, and therefore every soldier's first
    duty.
    Marshal Ferdinand Foch, *Principles of War*
    (1920)

11 Triumph cannot help being cruel.
    Jose Ortega y Gasset, *Notes on the Novel*
    (1925)

12 The problems of victory are more agreeable
    than those of defeat, but are no less
    pressing.
    Winston S. Churchill, speech in the House
    of Commons, 1942

13 Victory is a moral, rather than a material
    effect.
    Captain Sir Basil Liddell Hart, *Thoughts on
    War* (1944)

14 The vengeful passions are uppermost in the
    hour of victory.
    Captain Sir Basil Liddell Hart, *Defence of
    the West* (1950)

15 In war there is no substitute for victory.
    General Douglas MacArthur, address to
    Congress, 1951

AUTHOR INDEX

Note that in cases where authors have a topic section
to themselves, references to quotations by those
authors in their sections are set in **bold** type. Other
references are to sections and quotation numbers
within the sections.

Abrams, General Creighton 262:12
Admiralty message 31:17
Adams, Francis Lauderdale 145:4
Addison, Joseph 93:6
Aeschylus 229:3, 286:1, 340:1
Aguilers, Raymond of 172:3
Aitken, Kate 322:11
Akbar **1:1–2**
Alanbrooke, Field Marshal Viscount
    Alan **2:1–7**, 94:4
Alasdair, Mac Mhaighstir 39:4
Alderson, Sir Edwin 217:3
Alexander, Alexander 284:3
Alexander, Field Marshal Earl
    Harold **3:1–6**, 160:6, 174:4,
    228:8–9, 230:11, 235:13,
    278:16, 296:9, 14, 347:12
Alexander, Colonel Jeff 10:9
Alexander The Great **4:1–4**, 250:4
Alfonso XIII 52:7
Alfred The Great **5:1–2**
Ali, Mohammed 208:2
Alison, Sir Archibald 105:17, 210:4,
    5
Allenby, Field Marshal Viscount
    Edmund **6:1–7**, 55:10, 342:8
Allhausen, Major Desmond 274:2,
    283:11, 309:2
Allon 320:8
Alva, Duke Fernando Alvarez de
    Toledo **7:1–4**
Ambrose 119:6
Amis, Kingsley 200:17
Ammianus 250:6, 264:3
Aneirin 313:2, 326:2
Anglo-Saxon Chronicle 150:3
Anonymi Gesta Francorum 319:1
Anton, Quartermaster-Sergeant
    James 317:6
Archidamus, King 194:1, 258:4
Aristonice, Princess 196:4
Aristotle 315:2
Aristophanes 111:7, 8
Arkwright, Sir John Stanhope
    344:13
Arnold, Thomas 69:7, 198:7
Arrian 4:1, 6
Artabanus 196:1
Asquith, Margot 81:18
Asquith, Raymond 288:5
Astley, Sir Jacob 176:6, 319:3
Atahualpa 8:1
Attila The Hun **9:1–2**
Auchinleck, Field Marshal Sir
    Claude **10:1–7**, 86:9, 126:11
Auden, W.H. 255:14, 286:8
Austen, Jane 102:7
Austin, Richard 267:5
Aytoun, W.E. 63:5

Babur **11:1–3**, 212:5
Bacon, Francis 90:10, 234:1,
    300:2–3, 316:4
Baden-Powell, General Baron Robert
    12:1–9, 239:2, 320:5, 323:4
Bailey, Philip James 312:3
Baillie, Robert 85:2
Bain, Donald 335:8
Bairnsfather, Bruce 331:6
Baker, Sir Richard 265:3
Baker-Carr, C.D. **55:9**
Bakunin, Michael 266:3
Baldric of Dole, Bishop 241:2
Baldwin, Hanson 191:2
Baldwin, Stanley 274:7
Balian of Jerusalem 172:9
Banks, Sir Joseph 246:1
Barbour, John 122:2, 3, 165:1, 2,
    293:1
Baring, Sir Evelyn 62:10
Baring-Gould, Sabine 319:14
Barnes, Joshua 44:4
Barnett, Correlli 118:3, 154:12,
    161:13, 171:11, 190:3, 200:16,
    213:13, 236:12, 247:17
Barrault, Emile 244:7
Barrett Browning, Elizabeth 276:3
Barrett, Micéal 236:10
Barrie, Sir James 282:12
Bartlett, Ashmead 199:3
Bartlett, Vernon 73:12
Bartolomeo del Pozzo, Fra 240:5
Bashan, Raphael 86:8
Basin, Thomas 183:5
Bean, C.E.W. 215:5
Beaumont, Sir Francis 325:3
Beaverbrook, Lord 31:16, 19
Beck, General Ludwig 308:5
Bee, Brigadier-General Barnard
    75:4
Beecher, Henry Ward 266:4
Beers, Ethel L. 268:5
Beha al-Din 119:4, 172:11
Bel, Jean le 183:1
Belden, Jack 292:9
Belden, Jack 219:5
Bell, George 277:7
Belloc, Hilaire 93:10, 278:9, 302:3
Bence-Jones, Mark 127:5
Benedict the Pole 57:5
Bennett, Ernest 343:6
Bennett, Henry Holcomb 277:6
Bentley, E.C. 34:7
Beresford, Lt-General Viscount
    338:5
Bernadotte, Marshal Jean Baptiste
    13:1–3
Bernard of Clairvaux 241:3
Best, Geoffrey 323:12

Bethmann-Hollweg, Theobald 72:6,
    178:2
Biblical quotations,
    Epistles, I Corinthians 290:3, 307:2
    Epistles, I Macabees 4:7, 345:2
    Genesis 237:1
    Isaiah 314:1
    Jeremiah 250:5
    Job 307:1
    Judges 166:3–5, 237:2
    Kings 166:10
    Nahum 299:5
    New Testament, Matthew
        280:2, 311:1
    Numbers 166:1, 235:1, 304:1
    Psalms 166:11, 284:1, 316:1
    Samuel I 166:6–8, 237:3, 343:1
    Samuel II 166:9, 237:4, 339:1
    Song of Solomon 277:1
Bidwell, Shelford 66:7, 200:20,
    233:13, 238:9, 246:6, 257:16,
    261:16, 274:14, 308:10, 322:14
Bierce, Ambrose 283:10
Binyon, Laurence 344:9
Bishop, Billy 309:3
Bishop, Corporal Matthew 93:8,
    282:5
Bismarck, Count Otto von **14:1–3**,
    95:9, 228:4, 269:5
Blackader, Lt-Colonel John 186:2
Blake, William 255:9, 333:6
Blamey, General Sir Thomas 89:11,
    349:9
Blatchford, Robert 248:6, 325:9
Blind Harry 145:3
Bliss, General Tasker Howard 261:7
Bloch, I.S. 320:4, 330:2
Blockley, John J. 262:5
Blucher, Marshal 61:6, 130:5
Blucher, Prinz Gebhard Liebrecht
    von **15:1–3**
Blumenson, Martin 316:12
Blunden, Edmund 288:6
Boadicea, Queen **16:1**
Boece, Hector 264:4
Bogle, Eric 215:11
Bolivar, Simon **17:1–3**
Bonetti, Pascal 226:4
Bonham-Carter, Victor 124:9
Boru, King Brian **19:1**
Bosquet, Marshal 170:2
Boswell, James 248:2
Botha, Louis 349:5
Botwood, Ned 322:2
Bourne, Peter 304:8
Bradley, General Omar Nelson
    18:1–6, 234:10–11, 345:13
Brantome, Pierre de 240:3
Brasidas 194:2

Note that **emboldened** references are to entire topic
sections. Other references are to sections and
quotation numbers within the sections.